Bandwidth Bubble Bust

Bandwidth Bubble Bust

The rise and fall of the global telecom industry

Grahame Lynch

Authors Choice Press

San Jose New York Lincoln Shanghai

Bandwidth Bubble Bust
The rise and fall of the global telecom industry

Authors Choice Press
an imprint of iUniverse.com, Inc.

For information address:
iUniverse.com, Inc.
5220 S 16th, Ste. 200
Lincoln, NE 68512
www.iuniverse.com

ISBN: 0-595-18821-4

Printed in the United States of America

Dedications

In my eleven years covering telecommunications and business I have been fortunate to work with many fine and inspirational journalists and designers. These include Natalie Apostolou, Robert Clark, Tony Chan, John C. Tanner, Tim Marshall, David Binning, Dan Sweeney, Kirk Laughlin, Patrick Neighly, Martyn Warwick, Guy Daniels, Tony Whitehouse, Richard Chirgwin, Dan Tebbutt, Grant Butler, Alex Wills, Dick Wong, Shira Levine, Joan Engebretson, Paul DeVeaux, Mike Pickles, Lowell Tarling, Jaruwan Ngamman, William van Hefner and Rajesh Prothi.

I have also received generous support from Merle Lynch, David Haselhurst, Jeremy Grigg, Sally Lloyd, Danny Phillips, Sean Carr, Conor McCabe, Marc Spector, Cathryn van der Walt, Karen Dwarte, Peter le Gras, Marcus Cake, Mark Cramer-Roberts, Pip Bulbeck, Blake Murdoch and my wife, Chongko.

This book would not have been possible without the co-operation of hundreds of executives who have passed me information both on and off the record over the years.

This book is dedicated to all those who challenge the status quo and brave opposition in the pursuit of truth.

Contents

Foreword

This book grew out of nearly nine years of telecommunications journalism, mainly in Australia, China, Singapore and the United States. In that time, I've interviewed hundreds if not thousands of telecom executives, all of whom made promises, claims and statements of market superiority.

Few of the companies they worked for, ever lived up to their expectations.

When this was mixed with the investor exuberance of the late 90s, the result was a giant bubble, on the scale of the railway, tulip and automobile bubbles of history.

The massive deflation of that bubble in 2000 and 2001 should not be viewed as the mere consequence of a casino stock market.

Indeed, it is as much the result of failed execution, structural weaknesses in the global telecom industry and the confusing interventions of government.

This book tells the real story of the bandwidth bubble bust.

Introduction

Saturday, *March 31, 2001 was* just like any typical Los Angeles spring day. A little muggy and cloudy, but a lot warmer than most parts of the United States. The downtown area was as quiet and deserted as usual on weekends. With one exception. The crew from Qwest were in town, with their drills and arc welding equipment, laying fiber down First Street. That afternoon they were working a little up from the Los Angeles Times Building and City Hall. The local telecom revolution was in full swing.

A short distance up a hill, scores of bankers and business people were cursing. These were the LA types who eschewed the palm-lined streets of Westside for the plush condos of Bunker Hill, Downtown. They weren't happy. They were subscribers of the broadband services of Washington State-based CLEC ReFlex. Under exclusive arrangements with building managements, ReFlex provided discount broadband service. But not anymore. ReFlex had failed to secure an additional round of financing and had shut its doors. And to add insult to injury, its demise wasn't even newsworthy enough to make the newspapers.

L.A.'s newspapers were more pre-occupied with the demise of Northpoint, the nation's second largest CLEC. Having gone into Chapter 11, Verizon had entered into negotiations to take it over. These negotiations had failed. AT&T then picked up the pieces, but wanted Northpoint's co-location infrastructure, not its customers. Not any one of the 100,000 of them. They had their service cut off the previous week. And the day before, the California Public Utilities Commission had ordered that service for the company's Californian customers be restored for at least one month.

If ReFlex's customers were annoyed, they undoubtedly couldn't see the irony as they were forced to queue in the traffic jams outside their apartments—congestion caused by the Qwest installation.

For while Qwest was continuing with its nationwide fiber rollout, something very sick was happening in the so-called last mile. ReFlex and Northpoint were only the latest in a long line of competitive operators to suffer financial duress. The economics of the promised telecom revolution was failing. By the estimation of one industry association, just two of the country's four hundred or so competitive operators was profitable.

And this was in the United States, which had first introduced competition in the 1960s and had passed a milestone piece of telecommunications legislation in 1996 designed to set it in stone. On the back of this and America's formidable financing infrastructure, thousands of start-ups had secured billions of dollars in funding to make the equipment and sell the services to fuel that latest great American innovation, the Internet.

But the year 2000 ended the dream. It started optimistically enough with technology stocks surging to record highs. This was the new economy. The rules had changed. We were undergoing a new industrial revolution. Billion dollar valuations for companies without sales or profits were justified. A book came out predicting that the Dow Jones stock index would soon reach 36,000 points (one year later, it was trading at 10,000).

Even those that were skeptical about the prospects of dot-com companies, had to admit that the plumbing manufacturers—the Ciscos, the Qualcomms, the Lucents—had great prospects. At its height in early 2000, the stockmarket attributed a greater value to Cisco's stock than the entire annual value of the South Korean economy.

But then came the great tech stock crash. It started at different times for different segments of the market. For the big Internet suppliers such as Cisco, it started the last week of March. That week, Cisco stock had touched the $80 mark. One year later it was below $16. The wireless suppliers held on a little longer. Nokia's price peaked at $62 in June but had fallen to below $25 by the following February. The large incumbents were the last to fall. America's mightiest local operator, SBC, continued its climb until November, but then dropped back by nearly a third.

It took a while for reality to check in. Few analysts changed their ratings on stocks. A best-selling book released September was titled "Telecosm". In it, author George Gilder wrote "with total Internet traffic and bandwidth doubling every three or four months, markets have to learn how to evaluate Internet companies now facing less than one-tenth of one percent of the volume they can expect some five years hence." American fiber providers, unperturbed by their stock downturns, pressed on with their global expansion plans, confident that the competitive torrents unleashed by the 1997 World Trade Organization accord on telecom reform would soon be unfurled in Latin America, Europe and Asia.

But by early 2001, the telecom bubble had clearly burst. First came the layoffs. Global giants such as Motorola, Nortel and Ericsson announced massive staff cuts totaling tens of thousands of staff. Lucent even had its credit rating cut to junk bond status. Then came the service disruptions. Northpoint and ReFlex were to be the first of many.

The telecom bubble shared much in common with the aviation and railway bubbles of history. It is likely that the telecom industry will end up the same way, with relatively fewer players and low margins, even requiring some government subsidy.

This book explores how the bubble grew and burst, and examines the likely consequences.

It is based on nearly nine years of telecommunications reporting from over 20 countries with hundreds of industry executives and analysts. Many of those executives worked with companies such as Iridium, Project Oxygen, Telegroup and other outfits with grand plans that never came to pass. Others worked for major giants such as AT&T, BT, Hongkong Telecom and Telstra who are now faltering after years of assured profits and growth.

The bandwidth bubble didn't end in April 2000.

Indeed, it should have never started. Throughout the nineties, two phenomena characterized telecoms across the world. The first was the desire of governments to expose the sector to micro-economic reform in order to reduce prices to end consumers. The second was the increase in telecommunications equipment capability, to the extent that in some areas, potential capacities were increasing by several hundred percent between product cycles.

The explosion of the Internet as a business phenomenon blindsided both the telecom industry and its financiers.

The true story was that a combination of government-promoted competition and regulation was killing margins, while technology improvements were overwhelming demand growth.

This book is about how it happened—and how many in the industry still deny that it's happening.

Chapter 1

The submarine cable bubble

Although many of the headlines of the late 90s Internet boom focused on dotcom media and retail start-ups, a similar explosion of entrepreneurial activity took place in the global telecom space.

The old order of AT&T, MCI, BT, KDD and other trans-national telecom operators suddenly found themselves facing brash and effective start-ups.

Gary Winnick founded Global Crossing, raising funds to construct some 160,000 kilometres of fiber cable connecting 200 cities worldwide.

James Crowe took his executive team with him when his Metropolitan Fibre Systems company was sold to Worldcom and formed Level 3—using railway rights-of-way and the revenues of some profitable coal mines to roll out his own network connecting 150 cities across the Northern Hemisphere.

And former Microsoft executive Greg Maffei followed suit in 2000, with his Bermuda-based 360Networks unfurling its own plans to connect 100 cities worldwide, including many in Latin America.

The opportunities for these three operators, along with others with more modest regional ambitions, seemed limitless at first sight. Certainly, stock markets thought so, valuing the three of them at a combined value of over $100 billion in early 2000 even though they earned negligible revenues.

The international cable game had for years been an oligopoly of the PTT and monopoly carrier club. Carriers formed consortia to build cables that were planned out 25 years ahead. If you weren't in the club, you couldn't get access or only at an exorbitant retail price. With international voice calls priced high and no Internet demand, the regime was comfortable and uncontroversial.

But as de-regulation began to take hold outside of the United States in the mid-1990s, new competitive carriers began to find that they couldn't get access on essential cables. This, in turn, fuelled the opportunity for the likes of Global Crossing.

One market where this issue came to the fore was Australia, where liberalization began in earnest from 1993 and Internet adoption boomed from 1995. Telstra and its international club partners had a stranglehold on international connectivity into the country via the PacRimWest cable.

The major foreign shareholder of Telstra's first significant competitor Optus was none other than Cable & Wireless, the old British Empire carrier that had its own satellite and cable access to the country.

But new private competitors such as AAP Telecommunications, owned by a domestic news agency, couldn't get access. It eventually bought some access on a cable routed through Indonesia but it wasn't until last year—when the Southern Cross cable funded by Worldcom, Telecom New Zealand and Cable & Wireless launched—that a really plentiful source of international bandwidth to Australia was available to competitors.

The American advantage

The initial big opportunities for the global carriers were the trans-Pacific and trans-Atlantic routes, fuelled by the tremendous first-mover advantage enjoyed by the United States in Internet infrastructure and lowest-cost voice services.

With the explosion of ISP and IDD competition in Northern Asia and Western Europe, it seemed logical that the demand for inter-continental bandwidth would explode.

Global Crossing and Level 3 were quickly imitated. Some, such as Metromedia and Williams, saw opportunities in terrestrial fiber networks connecting scores of American cities and towns. Others, such as FLAG, went for a more global approach—connecting continents with each other via under-served routes such as the Middle East and Indian Ocean without actually connecting to hundreds of cities. And what was good for North America also looked good for Europe. Operators such as Viatel, Global Telesystems and COLT unfurled their own plans to create extensive pan-European networks.

One of these new generation entrepreneurs, Qwest, found that the late 90s stock market surge had given it the sheer stock currency power to swallow up its own US regional Bell company, US West.

Emboldened it then formed a joint venture with Dutch operator KPN to conquest Europe. Even after the 2000 stock market rout, Qwest's American and European holdings continued to be valued at an amazing $70 billion by the stock market. But, like any orgy, things got a little messy.

In 1998 and 1999, it was received wisdom that US Internet growth was exponential and that the emergence and growth of international markets would only accelerate this trend.

I remember John Sidgmore, the head of Worldcom's UUNet Internet subsidiary, travelling out to Hong Kong at the time to launch his company there and repeating his assertion that the Internet doubled in size every 90 days—in other words, a 800% annual growth.

Around the same time, Level 3 CEO James Crowe made similar assertions. Even former FCC chairman Reed Hundt gave credence to the claim in his 2000 memoirs.

A little later, the emergence of MP3 and streaming applications—which work best in a broadband environment—seemed to confirm the views of the optimists that if you build it, demand will quickly suck up supply.

In both January 1999 and January 2000, the common theme of presentations at an annual submarine cable conference held in Honolulu could be summed up as the following.

"We've never had any trouble selling capacity in the past, and the emergence of the Internet means that there's definitely no risk of that in the immediate future."

But these optimistic views ignored reality.

90 days is not a year

For a start, the Internet was not doubling in size every ninety days. One of the few telecom industry figures to challenge this assumption was Andrew Odlyzko, a mathematician at AT&T Research Labs in the US.

Odlyzko, who has published several papers on this subject, took a careful look at various indicators of Internet use, such as public data on academic Internet backbones and corporate releases from backbone providers and ISPs.

He reached the conclusion that Internet demand was more likely doubling every year. He couldn't find any significant network that was increasing in size by more than 150% annually.

Other data also suggested good reason for sobriety.

Nielsen NetRatings, a US Internet measurement firm, found that the number of "active Internet users" in the United States actually declined by some 10% or 7 million people between January and October 2000.

Later figures suggest that this number has dropped by another few million. Active Internet usage is defined as access at least once a week.

Additionally, an UK academic survey reported significant numbers of people who were "ex-Internet users", simply because they'd tried it, were bored or uninterested and had found better things to do with their time.

At the same time, the much-anticipated explosion in global broadband access networks failed to eventuate.

DSL and cable modem teething problems

In the United States, around 1 in 10 Internet users were connected to some form of digital subscriber line or cable modem service by early 2001. But across the rest of the world, DSL and cable modems stiffed. In Germany and the UK, broadband users only measured in tens of thousands, as dominant operators were slow to commercialize the service. Similar sluggish demand had been experienced in other advanced Internet markets such as Hong Kong and Australia.

DSL had even been problematic in the US. DSL involves a reconditioning of existing copper pairs to provide higher-speed data service. From an operator point of view, it is a nightmare to provision. Service quality can be impaired by line length or even unaccounted bridge taps on the lines.

Over-zealous marketing departments hyped up customer interest only to let them down through inadequate network provision and poorly organized technical support.

Things were even worse for the 25% of broadband customers who connected through competitive local exchange carriers. For the most part, these CLECs relied on Bells for technical provision of their service and although there were legal requirements designed to avert foul play, these were not terribly effective in practice.

This led to CLECs and Bells blaming each other for technical outages rather than solving problems along with all sorts of other unsavory behavior. By early 2001, many of these CLECs had hit the financial wall.

Sagging dotcom demand

Starved of the potential demand that could come from widely deployed DSL and cable modem access, the global broadband operators then suffered another problem—the dotcom crash.

One of the original motivators behind the global deployment of fiber was the expectation that booming West Coast US dot-com companies would take their business model to Asia and Europe.

Level 3 Asia CEO Steve Liddell told me that his company saw that market as the primary early adopters of their own Asian bandwidth service as he was setting up their operations in 1999.

But the dot-com crash of 2000 put paid to that. And at the same time, the American cost advantage in Internet infrastructure began to end.

Internet service providers—and the subsidiaries of these US dotcom companies—in early adopter markets such as Australia and Singapore quickly found that they could dramatically cut their major expense—international bandwidth—by creating their own web hosting and peering infrastructure. Previously, much domestic Internet traffic—even emails across cities—would travel across to the US because of the absence of this infrastructure.

In his research, AT&T's Andrew Odlyzko found that Telstra cut its US-Australia Internet link from 592 Mbps to 515 Mbps in 2000 as it developed its domestic Internet network and attracted global web-hosting providers to the country.

The year before, SE Asia's largest Internet service provider, Pacific Internet, even developed a novel scheme to share the cost of US Internet downloads across four different country markets by getting them to share the same satellite spot beam.

Indeed, as the US dollar appreciated to record levels against foreign currencies throughout 1999 and 2000, cash-strapped ISPs across the world had every reason to look to peering—as opposed to US dollar-denominated and linked bandwidth—as a cost-effective business strategy.

Corporate users outside the US were also proving circumspect in their purchase of broadband. As the editor of an Asian telecom magazine in 1998 I commissioned a survey of over a 1,000 major telecom users. That year they reported an average data usage of 64 kbps.

A year later it had doubled, as it did again in 2000. But tote up the combined data usage of these multinationals and conglomerates and it came to just one gigabit—far short of the hundreds of gigabits and terabits coming on line through new cable installations. In the latest survey, these users estimated that their demand growth was slowing from an annual doubling to something more in the range of 50%.

The triumph of technology

While demand fails to grow in the fashion that some anticipated, technology continues to develop at a fantastic rate.

Although many observers have become somewhat inured to the magic of fiber optics, the capacity gains of cables continues to accelerate.

There are three different parameters that affect the capacity of fiber cables and each one continues to double or quadruple with each generational change.

The first is data throughput, which has increased from a maximum of 2.5 Gbps to 10 Gbps. Suppliers say that imminent breakthroughs on chromatic dispersion should take that rate to 40 Gbps.

The second parameter is the number of lightwaves that can be supported on each strand of fiber—increasing from 16 to 32 and then 64.

The third is the number of fibers that can be sheathed in the one cable. The common standard of two pairs is rapidly giving way to four, eight and, potentially, sixteen.

Many of the new cables that have been installed over the past year support initial capacities of between 40 and 80 gigabits. But the newer generation, planned for installation from the end of 2001, are zooming into the terabit range.

360networks says its new cable connecting North and Latin America will support a maximum speed of well over one terabit—which the company somewhat ludicrously promotes as capable of transmitting 1.3 million digital photographs per second (or 40 trillion a year).

The latest flurry of announcements comes from India—these announcements seem to work on the basis of geographical fashion.

Singapore Telecom and Bharti Enterprises plan a South Asian link of some eight terabits. Not to be outdone, Tycom has announced a cable in the same region with the same capacity.

There's one major problem for these terabit cables. There's virtually no evidence of demand growth to support them.

The real demand curve

John Hibbard is a senior executive with Telstra, responsible for its international carrier business. He's renown for calling a spade a spade, even making it his personal mission at one time to single-handedly reform US Internet pricing. He told the Pacific Telecommunications Council in

Honolulu in January 2001 that at his best guess, there were just 100 gigabits of capacity actually being utilized across all international submarine cables.

His estimate seemed a little low given the assumptions and claims of the bandwidth barons. But at the same time, Global Crossing, the self-described dominant provider of IP backbone services in the world's second largest Internet market, Japan, had provisioned a trans-Pacific IP link of just 2.5 gigabits.

One of the first major multi-gigabit cables to service markets outside of the United States was the FLAG cable, which connected Europe with East Asia via the Middle East and South Asia. By late 2000, after three years of service, this cable had achieved just a 25% fill and, on the basis of FLAG's accounts, had yet to earn revenues sufficient to recover its capital costs.

Even if the total market continues to double every year, total global demand wouldn't reach 1.6 terabits until 2005. At the same time, there could be as much as 40 terabits of supply washing through the market.

This, plus the forces of competition, is obviously going to have a dramatic effect on prices. Indeed, it can already be observed.

In late 2000, US operator XO Communications (previously Nextlink) re-negotiated a bandwidth supply deal from Level 3 in Europe slashing a not inconsiderable $150 million—or about 47%—off the original price. Then in April 2001 it announced it was pulling out of Europe and had no need for Level 3's European capacity.

A survey by Pangea, which operates fiber networks in Europe, estimated that bandwidth prices in its markets would continue to decrease by 42% a year.

There's little on the horizon to indicate that prices will stabilize. There are some ten different cables vying for custom on the trans-Atlantic route and new ones continue to be announced.

Late in 2000, Cable & Wireless announced a new multi-terabit cable across the Atlantic called Apollo.

Rival Level 3 admitted that it was totally surprised by C&W's thinking. It was so concerned about the risk on that route that it sold down majority interests on its own cable to rivals Global Crossing and Viatel.

Price transparency
Competition and exponential technological advance aren't the only two drivers contributing to price cuts.

Another is increasing price visibility.

Until recently, there's been a vocal school of contrarian thought that refutes the idea that bandwidth is becoming a commodity. Epoch Partners' telecom analyst Mark Langner argued as recently as December 2000, that bandwidth lacks the key traits of commodities such as "liquidity, accessibility and price visibility."

An Australian Internet analyst, Ramin Marzbani, uses the more colourful metaphor of food. "There are different types of bandwidth as there are different types of food. You will always pay more for restaurant food or processed food than you pay for unprepared food," he told this writer in Thailand in 1999.

Since then, bandwidth exchanges offering B2B services for Internet and voice capacity trades have become particularly active. Although they are not particularly successful in closing trades, they do provide the price visibility that is so important in establishing common benchmark prices.

Prior to their emergence, bandwidth pricing was a highly mysterious, arcane and erratic practice. Only a few years ago, a 2 Mbps link from Australia to the US was priced as much as nine times as higher as the equivalent link bought in the US.

But bandwidth exchanges such as Arbinet, Ratexchange, Band-X and Asia Capacity Exchange happily post price bids on web sites. Almost every operator in the world has registered with these services and are obviously using them as information gathering and negotiating tools. One major buyer of capacity, a Hong Kong dot-com company, says it has a bandwidth agreement that calls for an automatic price reduction of 50% every year and the ability to re-negotiate this on a monthly basis.

In other places, bandwidth prices are becoming exposed to scrutiny as a matter of national mandate. Thailand's monopoly international provider CAT was to be the subject of a government audit, according to a March 2001 edition of Far Eastern Economic Review, because of concerns that its international leased line prices contained too much margin. Many other countries, concerned that business is being left behind by high bandwidth prices, are engaging in similar policy debates and public discussions.

Race for time

There's an obvious problem for an industry where volumes are doubling but prices are halving every year. Dollar volumes don't grow by nearly as much as is needed for all new entrants to achieve viability. Indeed, dollar volumes might not grow at all.

A case in point is Asia Global Crossing. It announced late last year that while it had only sold 3% of capacity on its trans-Pacific cable, this had returned 18% of its costs. Since then it has signed an unspecified deal with a Japanese operator as well as a deal for over 600 Mbps on a separate pan-Asian cable with Cable & Wireless.

But on the basis of an annual price deterioration of 50%, a further 3% sale of capacity in 2H 2001 would only return 9% of costs. The year after it would return just 4.5% of costs. And the year after, that capacity would sell at a loss.

In other words, Asia Global Crossing has little more than a two year window to sell capacity at a decent margin, in an environment where buyers are fully aware of price trends and loath to commit to current prices for more than a year or two out.

The CEO of Asia Global Crossing—which is owned by its namesake American parent along with Microsoft and Japanese dotcom incubator Softbank—John Legere, is aware of this risk, telling Hong Kong-based telecom commentator John C. Tanner in late 2000 that his company's first-mover advantage would prove to be a key to its eventual success.

Maybe this will be true. But if he is correct, the situation looks poor for global operators such as TyCom and 360Networks who are late into the party.

360Networks confirmed this in early 2001 when it issued a downward revision of its 2001 forecasts. The company said that revenues would fall 25%—or $900 million—short of expectation, although it attributed this more to the dot-com slowdown than a bandwidth glut.

FLAG, which was even later to the party in announcing trans-Pacific and trans-Atlantic routes, is experiencing similar problems. By early 2001, its stock was trading at a negative worth—its market capitalization was less than the company's cash on hand and about one third the value of its systems. By the time you read this, it's conceivable that the company may have fallen to a takeover attempt.

European rival GlobalTeleSystems missed a $12 million interest repayment in early January, having earlier sold out of its joint venture interest in FLAG's trans-Atlantic cable for just $135 million in cash.

Even Global Crossing may suffer some hiccups in the next couple of years. Its SEC filings contain a long list of the risks to its business. It's clear that Asia Global Crossing could suffer major business failure if minority shareholders Microsoft and Softbank fail to come good on promises to buy hundreds of millions dollars worth of capacity.

Contemplating failure

Given these disturbing trends it seems clear that not all major submarine cable ventures can survive the shakeout of the next couple of years.

On paper, the most vulnerable operator appears to be Viatel, which is building a network connecting 64 European and US cities. In early 2001, it told markets that its revenues would be below expectations because of a "temporary softening of broadband demand due to the reluctance of carriers, ISPs and ASPs to spend capital" and an overall slowdown in capital spending due to worsening economic conditions."

After having been valued as high as $3.5 billion in early 2000, Viatel's market capitalization had fallen to under $100 million a year on.

As the stock declined through 2000, its directors were heavy sellers. CEO Mike Mahoney alone sold stock worth $6.5 million during this period, mostly when the stock was trading at some 30 times its early 2001 levels. Viatel's SEC filings contain strong evidence of the effects of price deterioration in the sector—the value of its average billed minute fell 47% in the year to 3G 2000.

It's also likely that many planned systems or expansions to current systems will not eventuate.

The most high-profile failure was Project Oxygen, announced in 1997 by the former Nynex executive who dreamed up FLAG, Neil Tagare.

Oxygen originally intended to connect well over 100 countries on all continents with a mesh system that would have cost $10 billion in its initial incarnation. Even in the dot-com frenzy of the time, this was too much for markets to swallow, especially as the plan had no high-profile backers.

As time went on, Tagare scaled back his plans, as operators such as Global Crossing and Southern Cross made their own announcements for links on what would have been Oxygen's more marginal routes.

After a couple of years, Tagare had little to show for this effort. Former ITU secretary-general Pekka Tarjanne from Finland and an Ambassador Bradley Holmes from the US joined his cause, while a couple of minor international telcos such as Telecom Egypt, VSNL of India and Bezeq of Israel made financial pledges. When the project finally tanked in early 2001, its largest pledged backers were an Israeli medical imaging company and an Australian start-up carrier that had back-door listed through a speculative investment company.

Such are the capital demands of this industry that other operators are likely to need further financing to continue their rollouts. Markets are likely to refuse.

Dysfunction galore

At a Pacific Telecommunications Council conference in Honolulu in late January 2001, the just-retired Level 3 chief international executive Colin Williams surprised many when he rose from an audience to give a five-minute oratory that was startling for its candor.

Williams predicted that financial markets would not return billions of dollars raised for global cable rollouts. "We do not have a way to understand how demand is growing. We will have significant overcapacity," he said.

Up on the podium, industry executives who still had day jobs to return to were a little more cautious. But Telstra's John Hibbard also expressed concerns about a bandwidth glut.

And even more surprisingly, Jean Godeluck, the Frenchman who heads Alcatel's submarine cable division admitted "technology was outpacing demand".

Alcatel is one major winner from the current boom in deployments. Alcatel and TyCom monopolize the global market for submarine cable deployment, with KDD and Fujitsu of Japan playing a lesser role in that region.

Such was the bandwidth frenzy of the late 90s, that in late 1999, TyCom decided to join its customers and build its own global fiber network spanning four continents. The idea was hardly original, borrowing heavily from the Oxygen and Global Crossing model.

Not without coincidence, Oxygen subsequently collapsed—its founder Neil Tagare was palpably angry at TyCom's strategy during a January 2000 conference—while Global Crossing actually took legal action, alleging that TyCom had effectively stolen its ideas. After another legal dispute, TyCom agreed to give IDT Europe free capacity on its network.

One year on, TyCom is looking like it might have been too clever by half. Its executives are forced to play a delicate balancing act between not alienating its cable network customers while justifying its own move into their arena. Unfortunately, TyCom's logic sounds a little faulty.

The idea is that as a supplier, TyCom can build its own network at a cost advantage because it has no supplier margin and access to the latest technology.

Carrier customers can either buy capacity on this network or contract with TyCom for their own turnkey network. Since then, only one major customer—Singapore Telecom's C2C—has contracted with the company for a complete network system.

TyCom's strategy looks risky. It has admitted that its own network may consume as much as 75% of its supply capacity, which could effectively mean that profits on its network construction business could drop to one-quarter of potential levels. It hopes to make this up with margins on capacity sales on a network that won't begin to return revenues until the end of this year and won't be complete for years after. All this in a market where prices are dropping 50% a year!

Chapter 2

Tightening the Local Loop

It's been a given in telecommunications policy debates that the connection to the customer is the ultimate network bottleneck. The company that owns that final copper or fiber link from the kerbside into the office, household or factory has a hold on the customer that, usually, can only be prised away by regulation.

Back in the early 90s there was a school of thought that consumers and businesses were so fed up with monopolies that they would quite happily take a connection from a second network.

A series of regulatory reforms in Western countries led many to believe that local competition would flourish and help drive the exponential demand increases that were anticipated by the bandwidth barons.

But the local competition experiment hasn't done so well.

The Optus experiment

One example I followed closely was Optus Communications in Australia. Optus received licensing in 1992 to compete against that country's nationally owned integrated telco, Telecom Australia (now called Telstra).

The policy, which originated from that country's civil service but received political backing after a spirited debate within the country's left-of-centre government, was to essentially create a Telecom clone that would offer a wide array of long distance, cellular and other telecom services.

The ensuing creation, Optus, featured a motley crew of shareholder interests—America's Bellsouth, Britain's Cable & Wireless, an Australian transport group and a bunch of Australian insurance and mutual funds. By all accounts, board meetings had a certain edge. The government threw in its loss-making national satellite as part of the deal. A popular Australian chief executive, Bob Mansfield, who had previously headed McDonald's in Australia, was brought in to head the company.

Telecom would initially be required to enable its exchanges to allow code-free pre-selection of Optus long distance services. The government pitched in by conducting a series of regional ballots of consumers—where customers were asked to vote between Telecom and Optus (I can't find a precedent for any other company in any other industry getting such a leg-up in acquiring customers).

The first ballot in the nation's capital of Canberra saw Telecom retain over 90% of its customers, but Optus subsequently did better in larger cities where there were more immigrants and residents from other parts of Australia. By targeting much of its marketing firepower on those cities, Optus ended up with between nearly 15% and 20% of long distance customers—many of them the highest-yielding in the market. As the ballots moved on into smaller towns, Optus lost interest and its "vote" dropped to below 5%.

Fightback

Telecom, a government owned carrier, fought back in the style of a political heavyweight. It threw millions into advertising and branding, changing its name to Telstra (one of those horrible Qantas-type acronyms that Australian companies are so fond of). Optus was surprisingly composed in the face of this.

Its own brand awareness shot from zero to the nineties overnight and the media—buoyed by lots of new competitive advertising—gave the telegenic Mansfield and Telstra's ethnic American ex-AT&T CEO Frank Blount plenty of airtime to sell their message.

But Telstra's heavily politicized culture (the chairman reported to the Communications Minister and staff were highly unionized), meant that its strategy was influenced by highly Machivellian scheming. One masterstroke was to create a series of wholesale discount tariffs in order to create an US-style reseller industry. Soon a whole bunch of noisy resellers—in some cases, owned by really awful cowboys and spivs—were trolling the suburban streets of Australia, signing up new customers.

This helped destablilize Optus. Much of its pitch was based on a perception of cheaper price offerings but the reality was that it priced itself just a fraction below Telstra's prices. Although Optus was building out its own trunk links, it still relied heavily on interconnect agreements with Telstra which were priced extraordinarily high by global standards.

Telstra's well paid and aggressive lobbyists and lawyers ensured that the government didn't do too much about it.

Optus was also hamstrung by high costs forced on it by its license requirements which emphasized employment, universal service and investment targets. The mean and lean resellers were often operating with little more than a telemarketing desk and an Excel database, while grabbing up some lucrative customers.

Telstra's object in facilitating resale was to discredit competition. There was still a residual attitude among many customers that Telstra was a known, reliable quality and the new competitors were promising smoke-and-mirrors.

Telstra was at pains to ensure that the billing data and technical support it gave resellers was slow, poor and unreliable enough that they ended up hamstrung by customer complaints. Inevitably, Telstra began to remove the wholesale plans, replacing them with more complicated arrangements that gave reduced margins. One by one, the resellers went broke or consolidated.

The visual telco
Early on, Mansfield saw the writing on the wall and realized there was little future in discount long distance.

One of his chief technical executives, Dr Wayne Nowland, began driving a plan to give Optus complete control of the customer by building its own customer access network. In these pre-Internet 1994 days, pure voice couldn't justify a second CAN. But cable television together with voice could.

Australia didn't yet have cable television and Optus could potentially develop a lucrative new franchise. Even better, reasoned the technical visionary Nowland, there was an exciting new technology emerging in the US. Hybrid fiber-coax cable could support both voice and video transmissions. The technical trick was to demarcate the spectrum so that voice occupied some of the lower slots on the cable. Multiplexers at the exchange streamed the transmissions off from the video head-end and the public telephone network. On paper, the strategy seemed brilliant. Optus seemingly had a real shot of subsuming Telstra's dominance. Regulatory permission was duly sought and granted, and an American cable television company, Continental Cablevision, was brought in as an investor.

However, there were two major problems. First, anything Optus could do Telstra could do better. Within days of getting wind of Optus' plans, Telstra mobilized its own forces and hit back with its own plans for a cable TV network. Given its deeper logistic capabilities, Telstra quickly wrested back Optus' first-mover advantage and actually began beating it on network reach.

This two-horse race led to the second problem. Optus was no longer in a position to monopolize content and thus, a bidding war developed. Telstra found a useful ally in Rupert Murdoch's Fox, which became a joint venture partner and content provider.

Optus scored its own coup, lining up deals with many of the bigger Hollywood studios. All this came at a price of hundreds of millions of dollars (to this day, it pays penalties to the studios for having undershot subscriber targets).

But the real crunch was to come over sport. Australians were conscious of the fact that their own Rupert Murdoch had gone to Britain, bought up the football television rights and forced Britons to pay for the privilege of watching their favorite clubs go around every week.

No exclusivity for the innovator

The government, with one eye to the concerns of the electorate and the other to the concerns of free-to-air TV moguls, subsequently decided to ban cable television providers from buying exclusive TV rights on major sporting events. It even compiled a list of these events—the "anti-siphoning list"—that not only included staples such as cricket and Australian Rules football, but also golf, tennis and other second-tier sports.

Murdoch—not known for letting the intent of government policy getting in his way—had his own plan. Two of the three Australian cities where cable rollouts were targeted were a little different to the rest of Australian Rules Football-mad Australia—they preferred the Northern English sport of Rugby League. Free-to-air broadcasts of the Sydney Rugby League competition topped ratings for the Sydney and Brisbane free-to-air network of Australian billionaire mogul Kerry Packer. Murdoch decided he wanted in.

His strategy was extraordinary. He effectively created his own Rugby League competition, bribing many Australian and English clubs to defect to his competition with promises of higher player and sponsorship payments. Murdoch's son Lachlan was soon fraternizing with rugby league players and posing for photos as he signed more and more stars across.

The traditional football forces in Australia, backed by Packer's and Optus' money fought back with their own bidding war. Consequently, Rugby league—the supposed revenue driver for Australian pay TV—effectively split into two inferior and loss-making competitions. Telstra's cable TV company Foxtel and Optus had precipitated a content war costing hundreds of millions of dollars for a sport that attracted just a couple of million of fans.

The war was unsustainable. Rugby league ratings dropped. Lifelong fans turned off in disgust. Traditional clubs, some 90 years old, were forced into mergers or simply forced off the field.

Meanwhile, the free-to-air TV networks became more profitable. And five years on, the two cable networks—forced by reality to end their costly war and adopt essentially identikit programming—had attracted less than 10% of Australian households to their service. Telstra and Murdoch could at least claim victory as spoilers.

Optus' failure to win with cable television was compounded by the early failure of its telephony service.

Back in 1995 when Optus envisaged the service, telephony over hybrid fiber-coax was still highly immature. To cushion its risk, it brought in two vendors from the USA, Motorola and ADC. Neither had much luck in making the technology work. US trade magazines of the time hinted at the problems the two were having.

Optus didn't help its cause by bringing in an acerbic advertising man, Geoffery Cousins, who had little experience in the logistical and technical challenges of launching a communication network, as the CEO of its broadband unit.

Nevertheless, Optus pressed ahead, launching its service in 1996 with the Prime Minister of Australia, John Howard, making the first call using the technology to Optus' first cable telephony customer, an elderly couple who lived in a leafy suburb about one mile from Optus' main exchange.

Then nothing. Few customers were signed and those that had complained of problems such as noisy lines, problems in receiving calls and difficulties in hanging up. One of the main issues for the technology was getting the powering levels right. Too low and the phone didn't ring. Too high and the interference levels affected voice quality.

Optus' pricing strategy was also lame. It offered local calls at an untimed rate of A$0.20, just 5c below Telstra's A$0.25c rate. Not only did Telstra hit back with its own discount rates, but Optus' extra fees and charges meant that it would be months before an Optus customer would save their first cent on the service.

What Internet?

Meanwhile, Optus' preoccupation with rugby league and local telephony distracted it from the real action in the market place—the Internet. Throughout 1995 and 1996, the Internet boomed in Australia much as it had in the United States. But Telstra and Optus were a little slow on the uptake.

The most successful early player in the Australian Internet was a start-up called OzEmail, founded by a computer magazine editor, Sean Howard. As editor, Howard had set-up an electronic mail service for readers. His publisher, seeing little future in it, effectively gave the service to him. A couple of years later it had become Australia's largest ISP and had sold out to MCI Worldcom, giving that company a significant position in the country.

There were other success stories at the grass roots. In 1993, four 23 year old Sydney men started what became Australia's first commercial ISP out of an inner-city apartment. Seven years on, the company—now called Davnet—is expanding into Hong Kong, Chicago and Manhattan and counts no less than Japanese behemoth NTT as a strategic shareholder.

By the end of 1995, Telstra quickly realized its mistake and made a big play for the Internet market, rapidly rising to number two in the market and subsequently, number one about two years ago. Optus thrashed around with no Internet strategy until 1998 when it bought out one of OzEmail's lesser competitors. Despite its multi-billion dollar HFC network, it had not yet mass marketed a cable modem service as we went to press.

Five years after its HFC network launched, Optus has attracted a little less than 400,000 customers to the cable TV and telephony service, or less than 5% of the total market for telephone lines in Australia. Telstra's Foxtel had about 700,000 customers on its cable TV service. Telstra was also offering both ADSL and cable modem services.

Getting out

As I write this, Cable & Wireless (which eventually bought out Optus' other shareholders including the uncomfortable Bellsouth) was preparing to offload Optus to Singapore Telecom, a cash-rich operator keen to diversify out of its small home city. Singapore Telecom isn't necessarily too wild about its new acquisition.

Singapore Telecom had dipped its toes into the Australian market twice before—first, with a wholly-owned subsidiary, and then with a minority stake in number three operator AAPT—and subsequently pulled out. Its only getting into Australia now after preferred markets such as Malaysia and Hong Kong shut it out for political reasons.

Singapore Telecom will be buying into a company that, nevertheless, is one of the few profitable integrated competitive carriers in the world. In the last half of 2000, Optus made about US$37 million in profits. But its revenue streams reflect the reality of competitive carriage—almost half its sales come from its national GSM cellular network.

Telstra—which despite competition is continuing to hand in record profits—has survived much better than anyone expected. Like Optus, its most significant revenue stream and momentum is driven by its mobile services. The fixed voice market that drove the rationale behind Optus' HFC strategy appears to be shrinking in the face of discounting and wireless substitution.

Although Australia's market was open to complete deregulation in 1997, Telstra's army of lawyers, lobbyists and technical experts did a successful job in holding off some of the fine print of competition.

Number portability didn't come about as smoothly or quickly as competitors hoped, while interconnect pricing dragged on as an issue until Australia's anti-trust regulator became involved. I still vividly remember one Australian Senate hearing where both Telstra and Optus sent so many lawyers that the chairman looked up and remarked "what's this? Optii?"

Nearly ten years into the Australian competition experiment, Telstra's gross profits are still bigger than Optus' entire revenues.

Hong Kong Phooey

Cable & Wireless was clever enough to realize that competition in the residential market is a zero-sum equation for profits. Not only is it pulling out of Australia, but it also dumped its so-called "jewel in the crown", Hongkong Telecom. Cable & Wireless even struggled with its telephony and cable TV operations in the UK, which have now been largely sold off.

Inspired by Australia's example in the mid-90s, Hong Kong's British administration brought in Australian telecom regulatory official Alex Arena to advise on its own planned de-regulation. Cable & Wireless enjoyed a monopoly on telecom services in the city, renewed as recently as the Thatcher years.

It was also wildly profitable, returning up to half its sales as contribution to Cable & Wireless bottom-line. This profit was largely fuelled by Hong Kong's artificially small boundaries which allowed the company to charge IDD rates to families and businesses calling the short distances across the border to neighboring Guangdong Province in China and the gambling enclave of Macau. Hong Kong's position as a hub for regional operating headquarters also gave Cable &Wireless a monopoly over lucrative corporate traffic.

Arena had a difficult job. In my experience, he seemed a worldlier regulator than most—other regulators I have met have tended to be excessively naïve about the realities of modern business and telecom. Arena even displayed an intimate knowledge of such ephemera as fill rates on South Asian submarine cables. (As a postscript, he is now a senior executive with Pacific Century Cyberworks).

His task was extraordinary. He was being asked to liberalize telecom in a city about to undergo an historic transfer of ownership from Britain to China. A city that had no history of anti-trust or micro-economic reform and that was economically controlled by a handful of local and British conglomerates.

Opening up

Arena's first move was to open up the cellular market. In two waves of licensing, the city ended up with eight operators, later consolidating down to six. He, and his successor, Anthony Wong can be proud of their achievements. As of early 2001, Hong Kong had shot to the number one spot in the world for cellular penetration, with nearly 75% penetration. If you consider that many of the remaining 25% are small children or elderly people, this is a stunning statistic.

OFTA's second major move was less successful. It decided to issue three new local loop licenses. The winners were all strong players—Hutchison, New T&T and New World. While not household names in the West, these three companies represented the mightiest of Hong Kong's property conglomerates, which dominated economic life in the city. These companies weren't just home builders—they were owners of significant parcels of Hong Kong's commercial and residential real estate. In Hutchison's case, it also owned Hong Kong's major port facilities and some of the largest retail chains in the city.

If the true bottleneck was the property boundary, then these companies should succeed, went the logic. Their telecom companies should have a natural advantage in signing up their property tenants away from the incumbent.

But in practice they struggled. Two years after licensing, they had signed up no more than two or three percent of the market between them. Part of the problem was self-induced—they clearly lacked operational and marketing expertise. But there was a greater reality at work. Hongkong Telecom subsidized its local access service with its IDD profits. Without rebalancing, there was no way that the three new competitors could beat Hongkong Telecom on price and achieve profitability.

Breaking C&W's hold

Fixed competition could only work if Hongkong Telecom's monopoly was removed. At first, the government did everything it could without actually breaking the license agreement. Resale was licensed and encouraged, with the sop that Hongkong Telecom maintained control over the international gateway. This was no deterrent. Something like 120 companies secured IDD resale licenses in a short time.

But then the Hong Kong government decided to go all the way. In 1998, negotiations took place to end the 25 year monopoly agreement signed in 1983. Cable & Wireless drove a hard bargain.

Not only did it extract nearly US$1 billion in compensation for the loss of future profits, but it also gained an agreement to effectively double local access fees over a staggered period.

By late 1999, Cable & Wireless decided that it was time to leave. Singapore Telecom was an initial suitor, but Chinese political pressure led to local wonder start-up Pacific Century Cyberworks buying the Hong Kong telco.

The media loved the story of a Chinese dotcom taking over the colonial ex-monopoly. Pacific Century Cyberworks' chairman was a 33 year old, Richard Li, who was none other than the son of the world's richest Chinese man—Hutchison chairman, Li Ka-Shing. At first, the media adulation allowed Richard a charmed entry to the market. His Hong Kong IPO raised billions and he found it easy to raise debt.

Things quickly changed.

As global telecom stocks tanked throughout the world in April 2000, so did Pacific Century Cyberworks. By the end of the year, its stock price had fallen below 10% of its year high. It was forced to sell down a lot of his assets—for example, 60% of his cellular network—to Telstra, to help pay his debts.

And Hongkong Telecom handed in a very disturbing result in early 2001: a dramatic fall in sales and profits.

But Hongkong Telecom's losses are not anybody else's gains.

The Hong Kong telecom market is so bloodied that there may be no winners. The cellular networks are also struggling to make profit. A major IDD competitor, City Telecom, is ailing.

Hutchison's local network unit has formed a joint venture with Global Crossing, while Level 3 is now selling bandwidth. Although there is a temporary international bandwidth shortage into the city, new cable deployments may lead to a glut as demand growth falters.

Hongkong Telecom's efforts to develop a mass-market broadband access network have been unsuccessful. Using a fiber-copper hybrid, its VDSL network supports both a revolutionary interactive TV service and a high-speed Internet service.

Neither performed to expectation and there is talk that the TV service may be shut down as a cost-cutting measure. Plans by rivals to establish their own mass interactive TV and broadband services have not yet been realized.

The competition experiment has proved fruitful for Hong Kong's consumers, even if they had to subsidize some of it with their taxes in the form of compensation to Cable & Wireless. But low prices will only mean something if they are economically sustainable.

This, ultimately, is the problem of local competition.

Moving from a monopoly environment to a competitive environment is deeply disruptive. Breaking monopolies involves political sops such as compensation and requirements for new entrants. Newer competitors find themselves stymied by high wholesale pricing and volatile retail environments. Governments have to balance all sorts of contradictory interests in making policy.

Great ideas can become unstuck by the limitations of technology and finance markets.

It's not easy. But, as we know from the experience of utilities, governments will continue to distort the free hand of the market because telecom is viewed as too integral to daily life and economic activity to be left alone.

The USA experience
Although my two examples have come from the relatively minor economies of Australia and Hong Kong, they find echoes in the recent experience of the United States.

The US Telecommunications Act of 1996, although very much a result of political compromise, was aimed at breaking the natural monopoly of telecom companies and allowing local and long-distance players to compete against each other. The Act, along with further clarifications, also mandated the unbundling of the local loop so competitive local exchange carriers (CLECs) could get a leg-up into the market place.

Technological change was also leveling the playing field for competition. The development of microwave and millimeter wave technology provided the potential for operators to enter the local loop without the expense and delays of digging up roads and laying their own cable. America's regulator, the Federal Communications Commission, encouraged this development by allocating new spectrum in various bands above 2 GHz.

As was the case with Optus in Australia, US cable television providers also had new technological means to outfit their networks with voice capability and compete against incumbent local players.

Five years on, America's competitive environment is looking fragile. Of the 300 to 400 CLECs in the country, only two are profitable and many are going broke.

High-speed internet service provision is dominated by the Bell Operating Companies and cable providers, as CLECs and wireless broadband providers struggle to achieve critical mass. And as debt and poor execution cripple the main long-distance companies, the Bells are merging and strengthening their market position.

The CLEC sector is particularly ill.

CLECs were especially hit by the April 2000 stock market crash and some were trading at levels as low as 5% of their previous highs over ensuing months. By 1999, it was becoming apparent that few CLECs were making much money out of residential markets. In an analysis published in May 2000, my American colleague Joan Engebretson examined cost-based wholesale and retail rates across 39 states and couldn't find one state where BOCs offered residential access at a profit.

Her conclusion was that residential access was clearly subsidized by business customers and value-added services. CLECs, she opined, would be better off pursuing the high-speed data market.

Many of them were already taking that advice. Among the market leaders were operations such as Covad and Northpoint, who were gaining appreciable customer numbers in the DSL market.

But it seems the DSL market wasn't growing fast enough or sustainably enough for even the leaders to achieve profitability. As total US DSL deployments shot past one million in early 2000 and towards the two million mark, it became plain that it was still a technology in Beta stage.

Covad and Northpoint depended on Bell networks to deploy DSL and it's clear that the Bells were struggling with their rollouts. Consumer complaints with DSL rose to the fore as all sorts of technical and customer service problems emerged.

Companies such as Covad copped the worst of it. They would rely on information from the BOCs to install and service their customers but often that information was incorrect or delayed. In turn, major ISP customers of the CLECs were caught between a competitive squeeze and a service execution problem.

In late 2000, Covad revealed that 14 of its top ISP customers weren't paying their bills and 4 had entered Chapter 11.

Covad's revelation led to the departure of one of the US industry's brightest CEOs, Robert Knowling. An African-American and confidante of President Bill Clinton, Knowling had successfully raised $500 million for Covad just weeks before despite the harsh investment climate of the time.

Around the same time, Northpoint's largest reseller also entered Chapter 11 and it subsequently entered negotiations to sell out to super BOC, Verizon. After a good look, Verizon declined the opportunity— AT&T subsequently ended up buying it, although this was not a fantastic experience for the customers.

Rollback

Earlier, the FCC began a process of rolling back some of the regulatory advantages given to CLECs. In late 1999, the regulator exempted BOCs from unbundling requirements for network components such as packet switches and digital subscriber line access multiplexers—a move, which forced CLECs to spend more on infrastructure.

By early 2001, the FCC had announced an end to—or at least a cap on—reciprocal compensation. This was an arrangement that forced BOCs to pay ISPs and CLECs to accept calls originating on the PSTN for their dial-up Internet services. The new arrangement will see charges cut by over 80%, potentially denying CLECs between ten and twenty percent of their revenues.

The Congress has little political incentive to tilt the playing field in favor of CLECs. Many leading members who play an active role in telecom policy have received sizable campaign contributions from the big telcos.

US congressmen are increasingly swayed by the power of the dollar. Besides, the established telcos are big regional investors and employers. By contrast, there are few votes in successful micro-economic reform.

The FCC is little more supportive. Under the Democratic administration, William Kennard paid a lot of lip service to universal access issues but, in reality, pretty much passed every major media and telco merger that came his way.

The new Republican chairman, Michael Powell, is determined to stomp out what he terms "regulatory arbitrage" for special interests, which will probably make life even harder for the more vulnerable CLECs.

Business failure

However, the disappointments of US local competition can't just be sheeted home to government policy and powerful incumbents.

Competitive carriers are guilty of poor execution. One good example is the wireless broadband operators.

Although they theoretically provide service in almost every significant US metro market, they have failed to install networks that provide true point-to-multipoint capability and adequate coverage across cities. Many of the so-called wireless broadband operators are actually providing their services over leased DSL lines.

The three big long-distance companies—AT&T, Worldcom and Sprint—also failed to make much of an impression in local markets.

AT&T spent $100 billion on cable TV networks so it could provide killer bundles only to announce in early 2001 that it was changing the strategy and breaking itself up into four units. All three operators are also spending big on developing an obscure wireless broadband technology which they hope will give them cheap access into the local loop.

AT&T even has a name for this push—"Project Angel"—to which it apparently devotes hundreds of engineers. By March 2001, it had no more than ten thousand or so trial customers for the service.

But at least AT&T and Sprint have successful cellular networks to ease the growth pain.

Worldcom, which had looked unstoppable as recently as March 2000, was looking the sickest of the big US telcos by 2001. As voice migrated to wireless, Worldcom found that data wasn't growing fast enough to make up for its revenue and profit losses. In a make-or-break strategy, it attempted to buy Sprint, with an eye to its US CDMA network.

The takeover was aborted and with it went any immediate hope for Worldcom to get into the hottest consumer market in the US.

Cellular has proved the brightest spot of the US telecom industry.

After a slow start to digital adoption and some expensive spectrum auctions, the US penetration rate exceeded 40% or around 115 million users in early 2001. Consolidation saw the creation of six genuine national networks. Suggestions that spectrum ownership restrictions may soon be lifted could see further consolidations.

Although the US cellular industry isn't yet profitable, there are high hopes that the deployment of packet-based services in the next year will further push penetration and revenue rates. AT&T, for example, is working with NTT DoCoMo to deploy its popular I-Mode service in the US.

But it's the BOCs who enjoy the strongest position. Bellsouth, SBC, Verizon and Qwest (an honorary BOC courtesy of its US West buy-out) have largely seen off the CLEC insurgency.

With over 75% market share in DSL, they now face reduced competition and can slow down their capital spends to a more sustainable level.

They face no immediate serious challenges to their market position, and slowly but surely, are gaining regulatory approvals to move up the value chain into long distance.

The disappointment of competitive carriers across the world presents another problem for long-haul bandwidth providers. By Fortune magazine's estimates, they have installed some 36 million miles of fiber in the US alone.

Much of that investment was predicated on the promise of hundreds of CLECs supplying tens of millions of DSL lines and voice services. In turn, the CLECs were supposed to gain all the benefits of the dot-com boom and the resultant demand for the services of ASPs and ISPs. Dot-com is dead. The ISPs are going broke. The CLECs are following. There's a lot of toxin in the bandwidth food chain.

Chapter 3

Hubris in the sky

Looking back on the hundreds of billions of dollars of lost telecom stock value, one cannot help but ask how it happened in the first place.

How did so many large companies, intelligent people and conservative investors get it all so wrong? How could companies with solid revenue growth and high margins gain and lose as much as 95% of value in one year?

Of course, this sort of mass investment hysteria had happened before. Tulips, gold, nickel, railways, automobile and aviation have all experienced these types of bubbles. Telecommunications and the Internet was merely the latest manifestation.

The late 90s also saw an extraordinary convergence of unprecedented conditions for the industry. Obviously, the emergence of the Internet and multi-channel video as an economic opportunity was one. But we also had the WTO-mandated treaty on worldwide telecom liberalization. We had the emergence of technologies that enabled sophisticated wireless and data capabilities. We had the rise of financing mechanisms such as venture capital and second boards that encouraged entrepreneurial entry into an area traditionally funded by incumbent cashflow and government bonds.

Equally significantly, we had what I call the "CNBCization" of the technology economy—a process where CEOs of telecom and dot-com companies became celebrities and stock prices were as volatile and transient as Billboard chart listings. One couldn't pick up a business magazine such as Fortune without having a teenager or young executive such as Napster's

Shawn Fanning or AOL's Steve Case peering out from the cover—making all those executive readers in regular jobs feel inferior, old or both.

Then of course we had the dedicated tech-business magazines such as Red Herring, Business 2.0 and Fast Company, some of which were getting as thick as the Manhattan phone book by mid-2000. These magazines defined the dot-com cliché and contributed to the CNBCization of the industry. Red Herring publisher Tony Perkins was almost treated like royalty on his tours of global dot-com enclaves.

Fortunately, none of the major telcos I cover put any teenagers on the board although a few twenty-somethings dabbled in the sector—Lachlan Murdoch of One.Tel and Jason Ashton of Davnet spring to mind. But the great thing about the telecom bubble was that middle-aged men could also join in the fun. One of the most notable was Ted Pretty of Telstra.

Ted's pretty

A 1998 renovation of Telstra's management ranks saw two former Optus CEOs, Bob Mansfield and Ziggy Switkowski, appointed as chairman and CEO respectively. Despite the fact that both Mansfield and Switkowski had suffered inglorious exits from their Optus positions, the Australian government—which controls Telstra—thought that their competitive background would add a bit of zest to the place.

Switkowski certainly provided it. One of his first moves was to appoint lawyer Ted Pretty to his team. Journalists initially detected a certain level of contrivance when they noticed that Pretty shunned the suit and tie in favor of that ultimate middle-aged dot-com cliché—the black turtleneck.

Pretty then spent a lot of money taking expensive minority stakes in overvalued Australian dot-com firms with obviously modest prospects. After the 2000 crash, these investments tanked but it got even worse for Pretty.

A fellow who had been nabbed a few years before for securities fraud in North America ran one of these dogs and was duly exposed by a sharp young Australian business journalist, Nick Tabakoff. Pretty had apparently failed to run a background check during due diligence! Switkowski's response? He promoted Pretty to a position where he is responsible for 75% of Telstra's revenues. Millions of Telstra shareholders were dismayed.

The fact is that big blue-chip firms with checks and balances make big mistakes.

And it seems the bigger they are, the more the mistakes get a life of their own before anyone does anything about them.

A case in point is Motorola. This grand old American electronics company—at various times, the largest in its sector anywhere—managed to blow $5 billion on a global satellite network that almost every commentator predicted would fail.

The Iridium example is quite pertinent for explaining how the global bandwidth bust came about.

Legend has it that Iridium was born a decade ago when the wife of a Motorola executive was cruising at sea in the Caribbean and complained to her husband that she couldn't make a phone call. Motorola, which made wireless a specialty but at that stage sold nothing more sophisticated than an analog phone, proceeded to develop the idea for a global wireless phone, based on satellite transmission.

The original plan for Iridium called for 72 low-earth orbiting satellites. This was later "rationalized" to a mere 66. But such projects can't be realized overnight and, as a result, its development took nearly a decade.

A decade in the making

By 1991, Europe's telecoms vendors and standards bodies had developed their own GSM terrestrial cellular standard. Because of Europe's many international borders, the capability for network roaming was built into the specification from day one. European cell users took to GSM—and international roaming—with great gusto. By one estimate in 1999, GSM roaming was a $12 billion market accounting for as much as 25% of operator profits.

Motorola would have been aware of GSM's international roaming potential from 1990. It is doubtful that it would have anticipated the full extent of the future growth of GSM, but as GSM took off from around 1993, it certainly had plenty of time to make adjustments to its Iridium plans—after all Iridium didn't launch service until 1998.

But the Iridium story is one of denial. Tens of millions of dollars were spent on marketing and advertising, mostly many months ahead of launch. And despite the GSM success story, Iridium was marketed as a cell phone, albeit one that charged US$7 a minute and didn't work indoors. When the company collapsed in 1999, groaning under $5 billion of costs, it had 55,000 customers. London telecom commentator Martyn Warwick says he understands that 44,000 of them were never successfully billed.

In March 2001, Motorola was re-visited by the Iridium nightmare when creditors got American court permission to sue for $2 billion. According to wire reports at the time, they had a $47 million fighting fund to mount their case.

Iridium has yet to be put out of its misery. A VC firm picked up the abandoned network for a mere $25 million and the US Department of Defense, proving itself somewhat charitable, agreed to keep it afloat with $72 million of funding in return for 2 years of free use!

Motorola wasn't the only large American company to be blinded by the wonders of the space age.

Aerospace company Loral decided to launch its own global leosat network, Globalstar, at the comparatively bargain price of $3 billion. Globalstar's network isn't genuinely global. A user can only receive a signal if there is an earth station within the connecting satellite's footprint. This meant that the service launched with massive gaps in its coverage.

Into oblivion

Globalstar's decline began even quicker than Iridium's. Within months of launch, the company defaulted on shareholder loans. Things would have been even more grim if it wasn't for the fact that Loral and Globalstar share the same chairman, Bernie Schwarz. As one Globalstar executive told me, "Bernie will never let it go into Chapter 11." I guess it's all right just to punish all the other shareholders of Loral and its "partners" such as Qualcomm and Vodafone, instead.

Globalstar's target market—mindful of the Iridium experience—stayed away from the company in droves. Although the company had planned for 500,000 customers by the end of 2000, it ended up with only 30,000. In its early 2001 results, it declared a $3.8 billion loss on just a few million dollars of revenues.

Interestingly, Globalstar is marketed in a potentially more saleable way than Iridium. For example, Globalstar handsets can roam on to terrestrial networks, allowing users to access cheaper tariffs. The problem for Globalstar is that it is denied revenues for that usage and has little ability to actually measure what its customers are doing. The satellite mode is also a lot cheaper than Iridium's tariffs—around US$1 a minute versus as much as $7, although that is still well over ten times terrestrial equivalents.

Globalstar was also marketed a little more widely than Iridium. The Iridium service was pitched to the so-called International Man—the well-heeled global traveller. Globalstar quite happily sells anywhere. The company even claims that its phones are used by remote Canadian tribal communities!

Indeed, I believed Globalstar was a more attractive proposition until I met one of its senior executives in December 2000. But after a one hour discussion, I realized this company suffered from even greater delusions than Iridium. The said executive initially denied that global GSM roaming was its competitor and when pressed, suggested, erroneously, that global GSM roaming wasn't widely available.

He also blamed the media and financial community for Globalstar's poor sales by saying they had misleadingly compared it to Iridium (how could one not compare them?). And he then went on to blame the company's resellers for its woes suggesting that their sales people were too young and inexperienced! The only person not to blame, presumably, was himself!

But Iridium and Globalstar were in good company. Indeed, the story of the leosat industry is a 100% perfect failure rate. A wannabe Globalstar, called New ICO, entered bankruptcy before if even went aloft—it is now part of

Teledesic. And the cheapest of the new leosat networks, Orbcomm, also went into Chapter 11 in 2000 after it defaulted on $170 million of loans.

Of all the leosat networks, Orbcomm seemingly enjoyed the best economics. It eschews voice in favor of simple messaging—in essence, positioning it as a competitor to paging and dispatch networks.

Target markets included industrial applications in remote areas, transport networks, asset and infrastructure monitoring. On a visit to Japan in March 1999, I saw one customer who used Orbcomm services to monitor the location of its Tokyo-wide fleet of trucks, which carried dialysis supplies to home-bound kidney patients.

Subsequently, the company has entered the promising market of telematics, with Volvo agreeing to install the technology in all its new North American automobiles.

Orbcomm's limited feature set allowed considerable cost efficiencies. The network was launched in late 1998 with raised capital of about $500 million.

At launch, it had already shipped 30,000 units. But the company was clearly undercapitalized—it only had two key shareholders, Orbital Science and Teleglobe of Canada. With lower costs, Orbcomm still probably remains a better chance than Globalstar or Iridium for long-term survival.

Satellites can make money

With such an apparent pattern of failure, why does anyone even get into the satellite game? Big names such as Alcatel and Microsoft's Bill Gates continue with their own plans for multi-billion dollar constellations of satellites.

Well, others have made a success of satellite mobile services.

The best example is Inmarsat, a former co-operative of an eclectic group of national telcos that is now privatized and preparing for an IPO.

Inmarsat was formed to provide satellite communication services for the world's maritime community. Over time, its charter expanded to include land and aviation communications. It has provided this service for

20 years and thus, has an unenviable brand and reputation in the markets that it serves.

In its last reporting period, it made a $100 million profit on about $400 million of revenues with well over 200,000 terminals in operation. Its satellite network is made up of a mere 4 geostationary satellites. Tariffs aren't cheap, but at least one study suggests they compare favorably with GSM roaming and hotel charges.

Our leosat friends not only forgot about GSM roaming, but it seems they forgot about Inmarsat as well. The Globalstar executive mentioned earlier this chapter denied that Inmarsat was a credible competitor because "they were too expensive".

Globalstar's highest-profile fan, American stock pundit and futurologist George Gilder, devoted a chapter of his recent book Telecosm to Globalstar without once mentioning Inmarsat.

Now Inmarsat is upping the stakes with a considered move into the high-speed data world. It has already launched its so-called Global Area Network, providing speeds of 64 kbps over its existing network.

It then plans to spend $700 million on launching three new satellites that will provide data speeds of up to 432 kbps—by 2004.

But don't expect Inmarsat's success to herald in a new era of satellite rationality. Just wait for Teledesic which counts Bill Gates, cellular billionaire Craig McCaw, Saudi Prince Alwaleed Bin Talal, and believe it or not, Motorola as backers.

Teledesic wants to provide so-called Internet-in-the-sky services at speeds of 64 Mbps (that is not a typo) using some 288 satellites. So far, some $1 billion has been pledged to the venture.

The final expected cost? Just a mere $10 billion. Motorola shareholders, beware!

Chapter 4

The Great CDMA Hype

Ask someone to name the biggest revolution in telecommunications over the past decade or so and many would reply "The Internet". But on sheer numbers, they are wrong. The biggest revolution has been the cellular phone.

In the United States alone, about 67 million people use the Internet once a week or more. By contrast, 114 million use mobile phones.

Internationally, the comparisons are even more extreme. Across Western Europe, cellular penetration rates typically exceed 50% of the population. Less than one-third this number qualify as active Internet users.

For many, the Internet represents the triumph of the market. Despite its governmental beginnings, its growth has occurred outside the mandate of official encouragement and institutional endorsement.

Cellular is something quite different. Although few predicted its current market success, much of the development of cellular has occurred as a deliberate result of government planning.

Some 75% of all the digital cellular phones in the world use a standard that was conceived in Europe throughout the 1980s and commercialized in 1991. This standard, Global System for Mobiles (GSM), was very much designed with the interests of vendors and the political interests of European unification and industry policy at heart.

GSM uses a time-coded radio air interface that provides a good quality signal but suffers capacity limitations.

This forced operators to buy more infrastructure as their networks became loaded. But the standard compensates for such costs in other ways. It provides for standardized billing and signaling systems, opening the way for such nifty features as international roaming. Handset intelligence is based in a removable smart card, called a Subscriber Identity Module, which makes it easier for users to change handsets and customize their settings. GSM also incorporates the capacity for users to originate and receive short text messages of up to 160 characters in length.

To many non-Europeans, GSM was a typical case of continental over-engineering.

In the US, the preferred standard was a time-based air interface (TDMA) that was simply grafted on to the existing network systems used by the so-called first generation analog networks. Japan went down its own idiosyncratic path, developing the Personal Handyphone Service (PHS) and Pacific Digital Cellular (PDC) standards.

But over-engineered or not, GSM had an advantage. Not only had the Europeans developed the most fully-featured standard on offer, but they also boasted the most globalized vendor channels.

Government spectrum managers worldwide rarely had the resources to conduct much of their own research on emerging technologies so they usually relied on vendor advice. And there was no shortage of European vendor representatives ready to offer advice to license radio spectrum for GSM networks. This advice was quite clearly heeded.

By 2001, there were over 400 GSM networks operating in 162 countries across the world. Just two major economies—Japan and South Korea—had resisted the push.

The San Diego juggernaut

With Europe and SE Asia embarking on the GSM path and the US pursuing the 1.5G technology of TDMA, there was an obvious opportunity for a major "disruption".

And so it came, from San Diego firm Qualcomm and its CDMA technology.

While TDMA used a timing mechanism for its radio interface, CDMA used a coding mechanism. Qualcomm and its promoters claimed that this offered capacity benefits some tens of times that of the capabilities of existing analog and TDMA-based networks. In the minds of many, CDMA was the next big thing.

CDMA was nothing new. It was originally patented by none other than actress Hedy Lamarr, who together with her husband, gained inspiration for the technique from observing the musical coding on piano rolls. From the 1950s on, CDMA developed into a staple US military technology.

Qualcomm's real achievement was to take these ideas and apply them as a viable cellular technology. Driven by founders Dr Irwin Jacobs and Andrew Viterbi (who are still involved with the company to this day), it demonstrated a system as early as 1989. Jacobs proved an effective propagandist for the technology, The ensuing technical debate between the CDMA and TDMA camps was quickly labeled the "cellular holy war."

However, Qualcomm had no manufacturing experience and the European vendors weren't interested in funding its intellectual property rights. At first, Motorola proved the only vendor able to push the technology to commercial deployment stage.

While GSM systems were entering service across Europe, Asia and Australia throughout 1993, the first CDMA system wasn't to make an appearance until two years later—in Hong Kong, and then a few weeks later, in California.

But only a few operators in the Americas committed to using the system. Early reports were negative. Networks didn't perform to expectation and handsets were in short supply. CDMA had a big voice, but few results to show for it.

The Korean advantage

CDMA's first big break occurred in South Korea occurred in 1996. Throughout the early 1990s, the South Korean government had studied ways to develop comparative advantage for its electronics manufacturing industry.

Correctly observing that European vendors had the GSM market locked up, the government decided that the country's *chaebol* should pursue the CDMA path instead. The American CDMA forces, hungry for export success, eagerly licensed them the technology.

By all accounts, Korea's main wireless operator SK Telecom wasn't too happy at being forced to deploy a CDMA solution supplied by a Korean vendor under US licensing. With the inevitable delays of such an undertaking, SK Telecom wasn't able to launch a CDMA network until 1996.

The CDMA forces proved a little too guilty of believing their own propaganda. The early technical performance of CDMA was dreadful. One credible independent study showed it up as the worst of the three major standards. Conducted by a technical department of Hong Kong University in 1997, the study measured the performance of Hutchison's CDMA network against Pacific Link's TDMA network, and some six GSM networks operating in the 900 MHz band and the newer, immature 1800 MHz band.

The CDMA network was easily the worst for call dropouts, losing as many as ten times many calls as the better performing GSM networks. Vendors were slow to fix these problems because they had little experience with the core IPR and Qualcomm was too removed from the problems to offer speedy solutions. A study conducted in Singapore in 2000 gave the same result—GSM performed much better than CDMA in the field.

The Qualcomm forces, bolstered by the incredibly hyperactive organizing efforts of Californian PR specialist Perry La Forge and his CDMA Development Group, pressed on with their propaganda claims throughout 1997 and 1998.

But as European vendors did better with their GSM standard, eventually cracking the Shangri-La of mass adoption in China, the CDMA forces decided to take the game to another level.

The China prize

The US State Department was conscripted to the cause, and the contemporary Secretary of State, Madeline Albright led the charge. Letters were dispatched to Europe complaining of that continent's restrictive trade practices in favoring GSM over CDMA.

But the real prize was China. The State Department realized that Chinese adoption of CDMA constituted an effective bargaining chip over that country's admission to the World Trade Organization. Throughout 1998 and 1999, the CDMA/WTO trade-off became a key discussion point among China's top leaders. American negotiators, including those representing the CDMA lobby, couldn't help themselves, regularly discussing the progress of talks to the likes of the Asian Wall Street Journal. Those of us in the Chinese telecom reporting fraternity (I was one at the time) marveled at the naiveté of American negotiators who were flagging their tactics and agendas through press leaks.

The Chinese, living up to reputation, played a wonderful game of ambiguous negotiation. Premier Zhu Rongji would make positive noises about CDMA adoption. Then another top official, such as Communications Minister Wu Jichuan, would state something to contradict him.

As part of the push, the US State Department began pressuring allies in the Asian region to also adopt CDMA—a novel form of "cellular encirclement". Several countries in SE Asia issued CDMA licenses to obscure and under-funded operators. At time of writing few had launched, and one in Singapore had closed down. Taiwan's government-owned Chunghwa Telecom also announced a CDMA network only to quietly withdraw its plans some time later.

The US strategy was revealed in Australia in 1998 when the Australian government issued a press release stating that Telstra would build a CDMA network, one day before Telstra's board was actually due to meet to consider its own decision. My Australian newsletter, Communications Day, reported at the time that Telstra had been offered a "virtually free"

TDMA network from Ericsson. It later emerged that the US State Department had lobbied the Australian Prime Minister to adopt CDMA.

The 2,600% year

The second big break for Qualcomm came as standards bodies across the world came together to decide on the replacement for GSM and TDMA.

Almost everyone agreed that CDMA provided the best air interface for the high-capacity needs of future services. Although there were many competing proposals vying to get acceptance from the International Telecommunication Union, almost all of them were based on CDMA. Qualcomm was adamant. All of them would have to pay royalties to Qualcomm. It seemed that Qualcomm would get its eight cents worth of every dollar generated by cellular equipment into the distant future.

CDMA also seemed like it was on the verge of conquering North America. A number of fast growing networks operated by the likes of Airtouch, Bell Atlantic and Sprint were using the technology. CDMA networks had also been licensed in the virgin markets of Latin America. Meanwhile, the Korean networks were setting growth records.

Combined with the expectation that China would adopt CDMA as a condition of WTO entry, Qualcomm's stock price soared. It soon became one of the most widely held stocks in the United States. The stock's value rose an incredible 26 times in 1999, ending the year at a value of $200 per share and a market capitalization of over $140 billion.

Subsequent announcements fuelled the juggernaut. Korean and Japanese CDMA operators appeared set to upgrade their networks to the next-generation CDMA2000 standard by 2001 or 2002, providing lucrative new royalties. There were also suggestions that TDMA operators in the US such as AT&T and SBC may also dump their networks in favor of CDMA.

The great crash of 2000

Then it all fell apart.

Despite the propaganda and perhaps because of it, the global vendor community had never quite bought into the idea of writing out eight percent of their sales in checks to Qualcomm.

Companies such as Ericsson, Nokia, and Alcatel worked on their own version of CDMA, at first given the uninspired name of W-CDMA, but later to be given various monikers such as Universal Mobile Telecom System (UMTS) and 3GSM. NTT DoCoMo, keen to influence and exploit the world mainstream, joined their efforts.

German vendor Siemens, piqued that its preferred variant of CDMA technology had been shunned by its European rivals, went to China and started working with a local telecommunications academy to develop Time Division Synchronous-CDMA. Another Chinese start-up, LinkAir, came to Silicon Valley and recruited one of the world's leading cellular scientists, former Airtouch executive Dr William Lee, to promote its own technology—Large Area Synchronous CDMA.

The object of all these efforts was pretty similar. Take the core properties of CDMA. Then tweak and enhance them so as to put as much technical difference as possible between them and the patents of Qualcomm.

The industry was sending a clear message—we don't want to make Qualcomm rich. And, what's more, we think we can make a better technology than Qualcomm can. In March 2000, the great tech stock collapse claimed Qualcomm. Its mighty fall, which wiped nearly $100 billion off its value, was so complete that by mid-year some were wondering if a big European vendor such as Nokia might step in and take it over.

The long decline

Indeed, as the year rolled on, it became clear that Qualcomm's prospects were not worthy of that giant market capitalization.

China continued to drag its feet on CDMA deployment. While it gave a smaller operator, Unicom, the right to deploy CDMA networks, at time of writing, there were only four small city nets in operation and one under construction. Tenders had been called again for a national rollout for the second time in two years.

At the Asia Telecom ITU show in Hong Kong in December 2000, China made its true intentions clear.

China's communications minister Wu Jichuan stated that his country would not pay foreigners for the intellectual property rights used in domestically produced and deployed telecommunications technology. To accentuate his point, uttered in Mandarin, he said that while many people may have different opinions, the only one that counted was his.

Down the hall, Qualcomm CEO Dr Irwin Jacobs told a closed circuit TV service that technology transfers of CDMA to China were going to plan and that China would have to pay royalties on its homegrown TD-SCDMA technology.

Indeed, despite Jacobs' insistence on the patent issue, other vendors continue to maintain that they will have to pay little or no IPR on their own CDMA technologies. For example, many other vendors claimed to have their own IPR for W-CDMA, creating the potential for further legal disputes down the track.

Marginalized

As time went on, Jacobs seemed increasingly marginalized from the world's industry. In early 2001, AT&T Wireless ignored the pressure to adopt American-style CDMA and elected to go with the European GSM/GPRS standard, and ultimately, W-CDMA. Again, Qualcomm put on a brave face, issuing a press release that forecast lots of royalties but by now analysts and shareholders were becoming increasingly skeptical.

Even the CDMA hotbed of South Korea looked shaky for Qualcomm, with its cellular operators confirming that they, too, preferred to go the European/Japanese route for their future network upgrades.

In February 2001, Jacobs ventured into enemy territory—a GSM congress in Cannes, France—and predicted that W-CDMA deployments would be delayed by two years, providing an advantage for Qualcomm's own technology. Markets were unimpressed, wiping 20% off Qualcomm's value for Jacobs' act of talking down the market. Despairing investors would have preferred him to kept his mouth shut.

It's not even clear if Qualcomm's next-generation CDMA2000 technology (also called 1X) will beat W-CDMA to market. A claimed 3G network in South Korea operates at speeds of below 154k—and thus, is more analogous to a interim standard such as GPRS than a fully-fledged 3G net.

There were at least two W-CDMA trial deployments—in Britain and Japan—planned for the first half of 2001, and European operators who paid billions for new spectrum have incredible incentives to roll out their networks.

Ericsson insiders told me that the company is overwhelmed by orders for W-CDMA, which seems to provide it and others with every financial incentive to get the technology shipped.

By contrast, CDMA Development Group chief Perry la Forge told me in October 2000 that the process of balloting members to define the 1X CDMA upgrade would continue well into 2001.

And when it is completed, it may not follow Qualcomm's exact specifications because Lucent, Motorola and other vendors are also shaping its development. Few operators have even committed to 1X at this stage, further reducing the incentives for vendors to rush technology to market.

CDMA's big mistake

The big mistake of Qualcomm and its friends was to under-estimate the power of GSM and its ability to give operators what they wanted.

The selling point of CDMA is its capacity levels, which allow operators to support more users on existing spectrum and to more easily upgrade to higher data speeds, if and when demand for this eventuates.

But the selling point of GSM is its rich selection of standardized features that provide real incremental revenues. Capabilities such as SIM cards, short messaging and roaming increase customer value and operator revenues. When standardized across countries, GSM also benefits from tremendous economies-of-scale (it outsells CDMA by a factor of 4-to-1). This means that GSM handsets continue to benefit from increased range and superior form factors.

Qualcomm's view of the market was colored by its American origins. Like its TDMA counterpart, it rather lazily grafted its CDMA interface on to the old analog signaling standard. It also downplayed the need for solid, standardized feature sets.

When Qualcomm and other CDMA vendors finally realized the advantages of GSM's features, their reaction was too little, too late.

Qualcomm's prospective CDMA customer in China, Unicom, had been stating publicly for two years that it wouldn't adopt CDMA until the platform supported GSM SIM cards that would allow it to churn across existing GSM customers with ease. But it wasn't until late 2000 that such a solution was announced—at the joint initiative of Unicom and smart card specialist Gemplus.

A similar initiative to create roaming was finally developed by Asian operators in 1999, seven years after it became commonplace in GSM. This was a particularly grave failure. GSM networks generated as many as 8 billion international roaming calls in the year 2000.

In markets where CDMA competes against GSM, CDMA often comes off second-best. India's MTNL, after a careful evaluation of CDMA's lack of features, was moved to brand its planned service as the "poor man's phone". In Singapore, the technology was offered as a discount service and, then, simply withdrawn when regulators requested that the network shift frequencies.

Jacobs' capacity for hype is one thing yet to change.

In early 2001, Qualcomm forecast it would sell 90 million chipsets for the year. After seven years of commercial operation, CDMA had less than 90 million customers in total worldwide. CDMA adoption rates would have to accelerate to record levels for Qualcomm to meet its estimates.

By contrast, GSM had reached 452 million people globally by early 2001—about one handset for every 12 people on earth. Some 200 billion data messages are forecast to be sent to and from GSM handsets in 2001—40 for every human on earth. By the middle of 2000, GSM users were generating some 20 million international roaming calls a day.

Between Sep 2000 and Jan 2001, GSM added more users than CDMA had gained in its first six and a half years. Despite this, the CDMA Development Group continued to promote its platform as the world's fastest growing wireless technology.

In April 2001, Qualcomm stock enjoyed a P/E ratio of 455. The top GSM vendor (and likely to stay that way for W-CDMA), Ericsson, enjoyed a P/E ratio of 33.

Go figure.

Chapter 5

The Gilder Effect and other muddles

Over the years, Qualcomm has enjoyed much support from the analyst community, but few have been as enthusiastic as George Gilder.

For non-Americans, the attraction of Gilder is a little difficult to understand. He's an elderly man with a weedy high-pitched voice. Across his extensive career in the limelight, he has attracted attention as an anti-feminist baiter, a Richard Nixon speechwriter and later, a Ronald Reaganite-supply sider. His writing is florid in a classical way and he's a "big man historian", preferring to write about his subjects through the framework of epic struggle and heroic deeds from unlikely men.

But he gained his real fame—and money—as a self-styled technology pundit. Along the way he made many of his followers extremely rich. Gilder's reputation as a tech guru began with his 1989 book Microcosm. But it was his Gilder Technology Report—a slight 8-page monthly newsletter—that put him on the map.

Across 1999, this newsletter captured the stockmarket zeitgeist with its advocacy of "vision" technology stocks such as Global Crossing and Qualcomm. Gilder disavowed the newsletter's purpose as a stock tip service, claiming a more rarified educational role for it. But it was clearly promoted as a stock newsletter, and earned its reputation that year as Qualcomm increased its value by an incredible 2,600 percent.

The Gilder effect

Gilder's newsletter proved a rip-roaring success. According to some reports, it attracted 65,000 subscribers paying $295 a year—generating $19 million revenue. Added to this was his lucrative sell-out annual conference at Lake Tahoe, California; several other niche newsletters, a column in Forbes, another book, "Telecosm", and a well-paid keynote speaker business. This guy was in demand!

But the 2000 stock crash exposed Gilder. One of his favorite stocks, Globalstar, sunk soon after he incorporated it in his "telecosmic" stock list. Qualcomm also crashed back down to earth, losing 60% of its value. Global Crossing, and his other favorite, 360Networks, suddenly looked vulnerable. And his "conservative" picks such as Motorola and Lucent, lost market share and cut jobs.

May 2000 proved an embarrassing month for Gilder. Soon after AT&T Wireless launched its IPO, he secured an editorial leader position in The Wall Street Journal, where under a joint by-line with Richard Vigilante (his 'publisher'), he railed against AT&T's TDMA technology platform. The essence of his argument that AT&T was facing a technology disaster because it hadn't deployed Qualcomm's CDMA—at one point, he even criticized AT&T in patriotic terms, suggesting it was damaging America's national competitiveness.

Unfortunately for Gilder, the article didn't declare that he owned hundreds of thousands of dollars worth of Qualcomm stock, in clear violation of the newspaper's disclosure policy. The paper sheepishly disclaimed the violation by stating that it was satisfied that he bought the Qualcomm stock as a result of his investigations.

Later, Gilder got himself into a embarrassing spat with someone who was previously a fellow traveler, Level 3 CEO James Crowe.

Gilder had dumped Level 3 as one of his "Telecosmic" company picks, replacing it with 360Networks, which he claimed was using a superior terabit technology.

Crowe struck back, pointing out that this technology would not even be available commercially for well over a year. But more embarrassingly for Gilder, Crowe revealed that he had been approached to speak at a Gilder conference and was then sent a sizable invoice as required payment for Level 3's participation at the show. Crowe angrily declined to pay the invoice, and then found his company dropped from Gilder's list a short time after. Gilder, of course, denied any connection between the two acts.

The protection analysts

Gilder could be dismissed as a sideshow if it wasn't for his influence. Technology companies line up to appear at his shows and get a mention in his writings, conscious of the so-called Gilder Effect that an endorsement can provide for their stock prices.

One company CEO admitted to US telecom writer Dan Sweeney that he "shelled out" tens of thousands of dollars to attend Gilder conferences every year in the hope that the pundit would mention his stock.

But Gilder is simply the most famous of a legion of telecom analysts who run what some critics have described as a protection racket across the industry.

Sweeney's research into this resulted in a seminal America's Network article on February 15, 2001 titled "Exposing the Analysts".

His basic thesis—analysts quite happily pump up projections and claims about companies if those companies play the game and pay thousands of dollars for the research. Don't play the game, and your company will be disparaged, or worse, ignored, in the research.

Sweeney found that some analysts even take stock options in return for writing positive analyses of some companies. The inflated numbers help boost the stock and both parties benefit.

Even as late as March 2001, major analyst firms were still disparaging the idea of a bandwidth glut. Typical of the woolly analysis was this special from Phillips: "In direct contrast to current market sentiment that a bandwidth glut prevails, there is actually a global bandwidth gap. The gap will restrict more than 87% of the potential offered load from global access networks at the end of 2001."

They call this excess inventory in most industries.

Stock pushers

Even worse are analysts who work for financial firms. Months after stocks crashed in April 2000, only a few of several hundred US analysts were issuing "sell" recommendations. When Susan Kalla of Bluestone Equities disparaged several telecom stocks in early 2001, it was considered newsworthy enough to earn her a dedicated feature in American financial weekly Barron's. Her novelty value came from her act of initiating coverage on five telecom stocks with sell ratings.

Analysts often got so carried way that they punished companies for merely performing to expectation. In January 2001, Nokia announced a 64% growth in handset sales, exactly as it had previously forecast. But optimistic analysts believed that they might come in another 6% higher. When Nokia underwhelmed them, the stock was dumped, losing 9% in value the next day.

In other cases, analysts became carried away with exuberance for glamour companies.

A case in point is Pacific Century Cyberworks, a Hong Kong company which rode the wave of the late 1999 dot-com bubble to become one of Asia's highest valued technology companies.

Local investors loved PCCW as it was chaired by Richard Li—the son of the world's richest Chinese man, Li Ka-Shing. Mom and dad investors snapped up PCCW stock, hoping that they could profit from the Li magic. The stock was driven up to a value of HK$26.35 by mid-February 2000. This gave it a market capitalization of US$38 billion—an incredible number for a company that then had made just $373m in cash investments in various dot-com ventures and had $2 billion of cash on hand for acquisitions.

Likewise, the media loved the young Li. On New Years' Eve 1999, Li threw a party in Hong Kong so lavish that it overshadowed official civic celebrations. American pop diva Whitney Houston and husband Bobby Brown were flown in to perform at a reported cost of $10 million. Hong Kong's media had found their Great Gatsby.

Li's ultimate act of chutzpah was his audacious and ultimately successful takeover bid for Cable & Wireless' Hongkong Telecom, one of the most profitable companies in Asia. Despite the potential for indigestion, major Hong Kong investment analysts continued to call bullish price targets for PCCW stock. Between February and August 2000, Merrill Lynch, Lehman, HSBC, Prudential Bache, IBS Warburg and SG Securities nominated price targets of between HK$21.54 to $35 for PCCW stock.

But the substitution of promise with reality had a devastating effect on PCCW's stock value.

As the scale of the financing to pay for the deal became apparent, coupled with the realization that the ex-monopoly telco was little more than a declining annuity, investors dropped the stock. There was plenty to sell. Cable & Wireless had taken some of its payment in PCCW stock and wanted out.

By March 2001, PCCW stock, inclusive of Hongkong Telecom, was trading for a mere HK$3.30. Richard Li's humiliation was only beginning. At a press conference in Hong Kong on 28 March, Li announced a HK$6.91 billion loss (US$886m) for his company.

Hongkong Telecom's sales dropped 7%, mainly as a result of a 33% drop in IDD income. Li was also caught out on a personal front—an International Herald Tribune reporter found he hadn't graduated from Stanford University despite claims to the contrary in PCCW's PR materials and subsidiary investment filings.

Hong Kong's analysts continued to get it wrong. Some admitted that the deterioration of PCCW's financial performance had taken them by surprise. But as David Webb, an independent advocate of corporate governance in Hong Kong, pointed out, the analysts had little choice but to stay silent on the HKT takeover during the lengthy acquisition period.

Webb wrote in early 2000, "If you've noticed a wall of silence from respected firms on this deal, that's because most of them are working on it and so are not allowed to comment.

"Warburg Dillon Read (part of UBS) and BOCI (part of Bank of China) are making the offer on behalf of PCCW, which also gets "strategic and general" advice from CSFB. Meanwhile HSBC, BOCI (Bank of China), BNP (owner of BNP Prime Peregrine) and Barclays are the leaders of the debt syndicate."

"Merrill Lynch and Greenhill & Co are advising C&W. Jardine Fleming is advising HKT. Salomon Smith Barney is advising Pacific Century Regional Developments."

Journalism 101

This is hardly a phenomenon unique to Hong Kong. A core problem for financial analysts is their employers' reliance on lucrative underwriting fees from the companies they cover.

When pressed, analysts admit that they will never dwell too much on negativity. Although this is nothing new, this reluctance to criticize was exacerbated by the increasingly high financial stakes of the past few years.

Even now IPO business is down, there is a new round of mergers and acquisitions to look forward to.

Another problem for analysts is that they tend to be young and lack a sense of historical context. Many of them genuinely bought into the line that the telecom revolution was something new where the normal rules of economics didn't apply. They tended to accept as a given that elasticity of demand would save the day even where prices were declining and competition was increasing. They didn't look too closely at profits and they had faith that new technologies almost always satisfy frustrated demand.

Not helping the levels of dis-information are the journalists covering the sector. Much of the business reporting of telecom is dictated by the hourly cycles of newswire and Internet media—which often leads to simplistic assessments and distorting angles. When telecom stocks were hot, everything was positive. Now they're cold, everything is framed in a negative angle.

Trade journalists working for weeklies and monthlies are little better. Too many of them tend to be apologists for the companies and the industry they cover—their magazines are advertiser-funded and they work in symbiotic relationships with the industry's PR and marketing operatives.

If the journalists don't write up the company line, advertising gets withdrawn and editorial budgets get cut, or worse, they lose their jobs. Often this pressure is subliminal—few publishers have the gall to directly influence coverage but instead will make allusions to advertiser complaints about "lack of access to editors". The current climate, which has seen an advertising contraction, will increase pressure on editors to produce uncontroversial coverage.

There are probably more dedicated journalists covering telecom in the US than the rest of the world combined. Those in the rest of the world often operate in even leaner advertising environments. In some places, they are simply corrupt. Vendors and major operators regularly put on junkets in the expectation of favorable coverage—I must admit that my career as a telecom journalist has allowed me to visit some of the nicest cities and hotels in the world, often at some advertiser's expense.

I remember one press conference held in Jakarta, Indonesia by the local subsidiary of a major US vendor where the Indonesian telecom journalists in attendance were given wads of cash for simply turning up. In another Asian city, a telecom journalists' association was set up with the key objective of wrangling free mobile phones for members.

In other cases, the quality of journalism is reduced by a simple lack of resources—publications will quite happily publish white papers or press releases simply to fill up space. Few journalists remain in the sector for long as pay rates tend to be higher in the PR and marketing worlds. The result is a lack of industry memory or long-term commitment to readers.

Of course, many journalists and analysts try hard to do the best they can. But the economic constraints of their positions and their own interpretive limitations often combine to let their audiences down.

Blame Al

After a year of declining stock prices, some analysts are still looking for convenient scapegoats to explain away the end of the bubble.

One of the more novel was George Gilder who wrote in his newsletter in March 2001 that the tech stock slump should be at least partly attributed to the US Federal Reserve and the Congress.

"As long as the Fed assumes that inflation is a threat while prices plummet everywhere, the economy will be in jeopardy and all asset values will have to adjust to real interest rates in the double digits," he wrote.

"As long as Congress and the president think tax cuts cost money, there will be no significant tax cuts. The result is a slump, and where it ends, nobody really knows."

Funny, but the US economy only accounts for 20% of the world economy. Look at the bad results for telco companies across the world and they are the result of things that have little to do with US monetary policy. Some examples: the high cost of 3G licenses in Europe, declining bandwidth prices on Pacific and Atlantic routes, slower-than-expected growth in broadband access connections and the declining profitability of blue-chip telcos in international markets.

Gilder saw the stock slump –which wiped over 70% off the value of the entire telecom sector in one year—as a small stumble.

The crash, he suggested, would shake out "flakier firms, technologies, and business models, leaving stronger survivors to lead a new phase of wealth creation that will leave the [Warren] Buffets in the dust."

"The collapse of 2000 and 2001 will seem a mere blip in a long run bonanza."

Hmmm. For another opinion, this writer turned to Dr Marc Faber, the famed Swiss economist and investment contrarian. I interviewed him in September 2000, some six months after tech stock prices collapsed.

Faber wasn't surprised by events.

"Telecom stocks have peaked and will sell at the PE ratios of traditional companies. They have similar attributes to the electricity utility companies boom in the 1920s. They bottomed out in 1941 and did not reach their 1920s values again until 1965!"

"Telecom stocks reached their peak in March and while they may rebound here or there they will not reach the same high for another ten or twenty years. Telecom stock is still over-valued. I think they will drop by another 50%."

Faber is best known for his "Doom, Gloom and Boom Report", a monthly newsletter which often expresses pessimism at commonly-held assumptions about Western markets while highlighting opportunities in developing markets such as Russia.

Faber, who spent much of his working life in Hong Kong and now lives in semi-retirement in northern Thailand, believes Chinese telecom manufacturers are the likely beneficiaries of current trends. For him, the patterns of global investment which saw North Asian firms become dominant in electronics and automobiles will benefit China for telecom and IT manufacturing.

Many of the more fancied players in the IT&T space are little more than commodity makers, according to Faber. The cell phone will become as commonplace and cheap as the calculator, he suggests, while even Cisco's much-fancied server and router business is nothing more than a glorified commodity line.

The value of voice traffic over both fixed and wireless networks will continue to decrease as IP telephony becomes established, suggests Faber. The lowest-cost network operators will gain the rewards.

In an excellent biography of Faber published in 1999, Hong Kong writer Nury Vittachi recounts a speech the economist gave in 1997 titled "The Rise and Fall of Great Cities'.

Faber spoke of the great trading cities of ancient times—places such as Turfan, Khotan, Kashgar, Bactra and Lagash. Few people these days have even heard of these places. They all disappeared off the map as a result of hubris and economic change.

His argument was clear to his audience, who as Hong Kong residents, understood the historical threats to their own wealth and way of life. Companies can rise and fall, as have great cities. Capitalism and change destroys as much wealth as it creates.

Chapter 6

Spectrum auctions & other market distortions

If one looks at the history of wealth creation—and, in particular, of precious things such as diamonds and gold—a clear factor behind asset appreciation is scarcity.

Scarcity is often an artificial phenomenon. South Africa's De Beers' control of the world diamond stockpile is a case in point.

But despite the seeming abundance of technology potential in telecommunications, perceived scarcity played a major role in the bandwidth bubble.

And who was the ultimate beneficiary? Government.

The best example of this is the spectrum auction.

Auctioning air

Spectrum has traditionally been managed by government agencies for the simple reason that it needs to be regulated in order to prevent users from interfering with each other.

In its simplest form, this requires a basic policing role. In one example from several years ago, Australia's spectrum management agency conducted some investigative work when a taxi radio dispatch network became subject to some mysterious interference. The culprit? A deep-fryer in a fast-food outlet which was emitting feral radio signals!

But the ability to select who can use what part of the spectrum is also a real source of political and economic power.

For many years, this was more latent than real. Major spectrum users were defense agencies, aviation bodies and television broadcasters. There was little demand for commercial spectrum allocations.

What changed this was the advent of commercial trunked radio and cellular networks. Spectrum allocation became an important part of economic policy. For the first few years of cellular, the US FCC only issued two cellular licenses per geographic area, which dramatically enhanced their value. It wasn't until 1994 that new competition was enabled by additional allocations.

For new wireless technologies which only worked at certain parts of the spectrum, allocation policy could make-or-break their prospects. LMDS, which only works in the 24-28 GHz range, would be still-born if spectrum managers had acted slowly in allocating spectrum for its use.

Spectrum tax

Wherever government has economic power, it soon flexes its taxation power. Economists view spectrum auctions as the most elegant way to decide who gets to use a public resource. Governments shouldn't be in the business of selecting winners, they suggested, for this led to patronage and corruption.

For their part, government treasuries agreed. Western middle classes were over-taxed and a spectrum auction that raised money from a handful of companies was an effective way to top up budgets without political downside.

However, early experiences of spectrum auctions were less-than-successful. One example came from Australia where the government decided in 1993 to auction off two satellite licenses for television provision. The auction system employed allowed the highest bidder a time period of several weeks to make good on their bid—if they didn't, their bid elapsed and the next highest bid was considered.

A Sydney computer salesman, Albert Hadid, who had no background in telecommunications or broadcasting, manipulated the system, lodging dozens of spuriously high bids of descending value. As each of his bids elapsed it would default downwards to the next bid—always his. Hadid then cleverly exploited his position to gain funding for the licenses from unsuccessful bidders, when he had no clear assets or backing of his own.

The United States also saw its share of auction mayhem. Between 1994 and 1996, the FCC conducted nine auctions of spectrum for so-called Personal Communications Services, an entirely American concept which was essentially what others would describe as normal digital cellular services operating at higher frequencies.

Unlike most countries, which simply divided spectrum into parcels and sold them on a national basis, the FCC went with a convoluted scheme designed to promote innovation and new competitive entry.

For a start, existing operators were restricted from buying new spectrum in areas where they already held spectrum.

But the FCC divided the country into 51 divisions and 493 basic trading areas. Two lots of 30 MHz spectrum was sold across each of the 51 divisions, with preference given to companies that had "contributed substantial innovations" to PCS technology.

Another lot of 30 MHz spectrum was sold across the 493 basic trading areas, restricted to smaller companies with no existing cellular operations. An additional three lots of 10 MHz spectrum were sold across these 493 areas.

One of the major winners was start-up NextWave, which spent $4.2 billion buying the "smaller companies" spectrum. It obtained support from Qualcomm, but this wasn't enough—the company missed its license payments and eventually succumbed to bankruptcy.

The fragmented nature of the PCS auctions made it difficult for companies to secure national footprints and claim the resulting economies of scale. Only one of the successful bidders, Sprint PCS, has built anything approaching a national network. Other regional operators using the GSM standard entered into mergers to form VoiceStream, in late 2000, but it remains a minor player in the market. None of the PCS players have come near returning their investments and the industry as a whole continues to lose money.

Meanwhile, pioneer and pro-competitive restrictions continue to distort auctions. The early 2001 PCS auctions saw AT&T pick up much of the spectrum –using a proxy company owned by Alaskan natives!

European excess

But the $30 billion spent by US operators for spectrum was chicken-feed next to the amounts bid by British and German operators in 2000 for so-called 3G spectrum.

3G spectrum was allocated in the 2.1 GHz band, above those frequencies used for conventional cellular and PCS uses. 3G is a concept promoted by vendors to describe the next-generation of cellular services, which would support data, video and Internet use as well as conventional voice.

Existing operators had two reasons to bid for the newly allocated spectrum—the new spectrum was necessary to make these services work effectively, and secondly, if they didn't get it, someone else would, which would increase their competition and reduce their profits.

Britain was the first to auction 3G spectrum and the intensity of the bidding took everyone by surprise. As prices escalated, bidders began to drop out. Eventually, the four existing cellular operators all picked up 3G licenses. The only new competitor was a Hong Kong/Canadian consortium, which picked up the fifth license.

The cost of all this? US$35 billion. Half was payable immediately and the remainder by 2006. The total amount that was bid equaled nearly $600 for each Briton. Or well over $1,200 for every existing cellular subscriber.

The British result was exceeded in Germany. With six licenses on offer, the total amount bid reached $44 billion. Like Britain, this was equivalent to nearly $600 for each German. Unlike the UK, German telco executives were honest in their public statements—the sums were too high and there would be difficulties in making their investments pay.

Investors were shocked. Stock prices of the major cellular players plummeted. Subsequent auctions were less speculative. Two licenses in the Netherlands went for just $2.6 billion. Six 3G licenses were let for just $600 million in Australia—a similar sized economy to the Netherlands. The market there valued 3G spectrum at about half the amount paid for newly allocated 2G spectrum sold off less than a year before. Given the uncertainties surrounding 3G, that was probably a fair economic assessment.

Other countries eschewed the auction route in favor of the beauty contest. Japan and Finland gave out licenses at virtually no cost. Many other countries followed suit. Having a bet both ways, France decided to allocate four licenses without auction—but slapped charges of over $4 billion a piece on them. So far, only two operators have taken up the offer. In Belgium, one of the four licenses on offer remains unwanted by the market.

The tax that dare not speak its name

The economic case for spectrum auctions focuses on the notion that radio spectrum is a public and natural resource. Thus, private companies should compensate the community for the profits they make on use of this resource.

But there's one problem with this idea. It has been applied to cellular companies in a highly discriminatory fashion.

Similar arguments apply to agricultural and aquatic industries. For example, economists suggest that the best way to solve the problems of over-fishing would be for countries to auction off permits for fishing rights. Fine in theory. But in practice, most of the world's agricultural and aquatic industries aren't compensating governments for their abuse of natural resources. To the contrary, many of them are actually the recipients of subsidies designed to win their votes.

If cellular companies are profiting from a cheap or free public resource, then there's two ways the public can benefit. The obvious first way is through the price benefits that come from lower costs and competition. Spectrum auctions distort competition because they create artificial scarcity. Governments have incentives to set licensing conditions and numbers in a way that maximizes government revenue.

The second way is to tax cellular companies on their use of spectrum when they actually start to earn revenues and profit. This principle seems fine for the rest of the economy and it's a great mystery to me why governments go on about the need to kick-start innovation and technology, and then proceed to bill the cellular sector for spectrum that they haven't even begun to earn revenues from. Some European countries clearly agree. Sweden will charge 3G operators a spectrum fee equivalent to 0.15% of annual revenues for the duration of the license. Spain's $3 billion spectrum fee for its four 3G operators will be levied in 20 annual installments.

Despite the perception that cellular companies own licenses to print money, it is a tough business from a margins perspective. After ten years of operation and 16 million customers, AT&T Wireless still does not turn a consistent profit. The world's largest cellular company, Vodafone, lost $5.92 billion in its last full reporting year, although that partially reflected the cost of one-off acquisitions.

If one takes the most generous measure of its performance—earnings before interest, tax, depreciation and amortization—it made $4.8 billion in earnings. But this is split between networks in some 25 countries—or less than $200 million per market. Or looked at another way, it had 39 million customers and was making a little over $100 on each one in raw profit. On a net basis, it ended up losing nearly $150 per customer!

Auctioning what you don't have
It gets worse. The US is now preparing to auction off 3G spectrum allocations already held by existing broadcasters. The FCC wants to auction off this spectrum—in the 700 MHz band—in 2002. One problem—the broadcasters don't have to vacate this spectrum until 2007 or until digital television signals reach 85% of US television households, whichever comes later. The 3G telcos are expected to compensate the

broadcasters for any early removal from the spectrum, presumably at whatever price the broadcasters nominate.

The FCC had already technically broken the law by delaying this auction to 2002! The Congress required this spectrum to be auctioned by the end of September 2000, even though the existing users have use of it for another six years.

At the same time, Verizon Wireless asked the FCC if it could delay the payment of $8.8 billion that it bid for PCS spectrum in 2001. The reason? NextWave, the original owner that had entered Chapter 11, said it had now come up with the money for its original bid for the spectrum some five years before. It has launched legal action to recover spectrum it believes it rightfully owns.

For some, these issues may appear esoteric—a simple conflict between companies that aren't particularly important in terms of the national interest. Cell phones, like digital television, are a luxury, they argue, and if the companies that provide them are the victims of wonky policy and unfair taxation then so be it.

But the same FCC does seem to regard the cell phone as something intrinsic to the national interest. In 2001, it is requiring all US cellular operators to install global positioning system (GPS) capabilities in cell phones and base stations as part of what it calls the E911 initiative. The idea is that emergency authorities should be able to pinpoint the location of distress calls made from cellphones—apparently, in 1999, over 40 million 911 calls were made over cellular networks. The industry isn't complaining too much, as the 911 requirement provides a kick-along to potentially lucrative location-based services. But such a policy initiative does raise the question of whether it is in the American public interest for the cellphone industry to be the subject of unfairly expensive and confused licensing requirements.

It's clear that European governments also see deployment of 3G services as a public interest issue. Most 3G licenses in Europe have rollout conditions attached. In Britain, each of the five 3G operators must offer services to 80% of the population by 2007. Other countries have imposed tighter deadlines. Both Germany and Austria have mandated 25% coverage within two years. Governments in Spain and Portugal are even more demanding. Portugal's four operators have to cover 20% of the nation by November. Spain's four operators have to offer 3G to all cities with more than 250,000 inhabitants by June 2002!

Grabbing hands

Of course, it is not just cellular operators who have cause for complaint about unfair government imposts.

The entire telecommunications industry is subject to complicated taxation and licensing arrangements. Take the case of the typical Pacific Bell customer in the western United States. He is compelled to pay at least ten different industry and government taxes on their phone bill—a Number Portability Service Charge, a California High Cost Fund Surcharge, a California Teleconnect Fund Surcharge, a Universal Lifeline Telephone Service Surcharge, a State Regulatory Fee, a California Relay Service and Communications Devices Fund charge, Federal Tax, 911 Tax and Local Tax. A typical phone customer in the United States might pay over $200 of special phone taxes a year. Then, of course, the BOCs and all other phone companies pay the standard range of business and corporate taxes.

Other countries aren't so liberal with these types of taxes and cross-subsidies, but the United States is by no means unusual. In some places, consumers have it even worse, because their dominant phone company is owned by government and is often required to pay special dividends to treasuries in addition to taxes.

Telstra paid out some 15 cents in every dollar of its sales last year to the Australian government in the form of taxes and dividends. It also paid out another 5 cents in the dollar to pay for universal service requirements and its bids for spectrum for the 1.8 GHz and 2.1 GHz bands. And it was still paying the opportunity cost for two government-imposed decisions—a mandate to extend ISDN provision to over 95% of the population and another mandate to close down its analog cellular network and replace it with a CDMA network.

As part of its license requirement, Telstra's CDMA network must cover over 96% of the population, as had the analog network. This is no mean feat in a country the size of Australia. Unfortunately for Telstra, the analog network's one-and-a-half million customers had mostly defected to GSM networks. The expensive new CDMA network has less than 300,000 customers.

The great cell sell-off

In early 2000, cellular operators enjoyed buoyant stock prices. The general April technology stock crash was just the beginning of the end. As results from the European spectrum auctions came in and investors began to evaluate the true timelines and costs of 3G deployment, cellular stocks fell further in late 2000 and early 2001. Vodafone, for example, lost over $160 billion in market capitalization since its 1Q 2000 high. As cellular operators stocks fell, so did the share prices of major wireless vendors such as Nokia and Ericsson.

Is it any wonder that cellular operators are among the most unfashionable companies in the world?

What other industry is forced to bid billions of dollars in order to gain the right to offer a service that has yet to come out of the laboratory? What other industry faces universal service targets that must be met within a matter of months for a technology that has yet to be commercially proven anywhere in the world?

3G may yet prove to be a success. Mobile data accounts for anywhere between ten percent and thirty percent of the revenues of the more cutting-edge cellular operators in Japan and Scandinavia.

But the European Union itself identified the shaky prospects for 3G in a January 2000 report. With classic bureaucratic understatement, it pointed that 3G will be "burdened by some very heavy front-end expenses". How much? 130 billion euros for government spectrum fees and at least as much again for actual equipment build-out. The cumulative investment to date for all 2G GSM networks in the EU? Just 70 billion euros.

That's a 130 billion euro premium for a "scarce" resource that's yet to return one dollar in revenue.

Chapter 7

Chasing China and other global follies

Shenzhen has 3 million residents. As a city, it is less than 20 years old. This southern Chinese city is the perfect example of the massive wave of industrialization and wealth creation that has swept China in recent times.

Just 20 years ago, Shenzhen was a fishing village on the wrong side of the Hong Kong border. But then China declared it a special economic zone, giving businesses located in the city many special privileges not afforded other provinces. The economy boomed. At first it became Hong Kong's manufacturing town. Now it has become an important hub for China's emerging technology export sector.

Two of the more interesting companies in Shenzhen happen to be telecommunications manufacturers. Huawei Technologies and Zhongxing Technologies, located across the road from each other, have come out of nowhere to become major players on the world stage.

Huawei was launched with US$1,000 in capital by four entrepreneurs a little over one decade ago. Zhongxing had slightly more official backing, gaining status as an official state-owned enterprise. After sixteen years of operation, it is now listed on a Chinese bourse.

For all the talk of exponential industry growth, Huawei and Zhongxing actually live it. Huawei's revenues increased from US$1.66b in 1999 to $2.5b in 2000. It expects revenues to surpass the $5b mark in 2001. Zhongxing is nipping at its heels, recording US$1.23b of revenues in 2000.

Huawei claims strong market positions in cutting-edge sectors where one would expect that advantages might accrue to foreign companies. For example, it claims a 30% market share in China's optical networking market and a 40% share of Internet access servers. In the Internet router market, it claims a 13% share, higher than global giant Cisco. For its part, Zhongxing claims as much as a 70% share of China's videoconferencing market.

Huawei says it will export some $1 billion of product this year. Some of it will take the form of OEM product for American vendors. Others are outright turn-key contracts to operators. Zhongxing will also export a significant proportion of its output to overseas markets, notably in several African and Central Asian countries where it has a surprisingly strong sales presence.

The China mirage

For foreign China bulls, the success of Huawei and Zhongxing constitutes clear evidence of the riches to be found in the Chinese telecom market. Their market position will be easily contested when China enters the World Trade Organization, suggest the bulls. Foreign vendors will be able to compete on a level playing field, and with normal intellectual property rules in place, repatriate the full value of their exports and manufactures.

China already boasts over 140 million fixed line connections and nearly 100 million cellular connections. But as the bulls like to point out, these penetration rates constitute only 10% or so of the population. With economic growth of 8% and rising incomes, there are still potentially hundreds of millions of new customers across the country.

More encouraging signs come from recent developments, say the bulls. China's dominant fixed and mobile carriers are pursuing international stock listings, increasing their access to capital and corporate transparency. China's first independent wireless operator, Unicom, seems to have finally found its feet, and is pursuing plans for a major network upgrade. And new competitive backbone operators such as China Netcom and Railway Telecom have also won official endorsement.

Foreigners certainly are enjoying the first signs of real access to mainland telecom investments. Vodafone paid some $2.2 billion for a mere 2% stake in China Mobile, a Hong Kong listed entity that operates dominant networks covering 48% of the Chinese population. Rupert Murdoch's News Corporation was part of a consortium that paid $325 million for a 12.5% stake in Netcom, which is building a 12,000km fiber network across the country.

But for those foreigners expecting to make their fortunes in China, recent developments are likely a case of too little, too late.

No help for the locals

Some China observers attribute the success of Huawei and Zhongxing to domestic industrial policy that rewards local companies. WTO accession will remove some of these advantages and help foreign companies, they say.

But such notions are not entirely correct. The most established and successful "indigenous" telecom manufacturers in China are foreign. Western vendors such as Ericsson, Siemens and Alcatel have been operating in China for decades. They have highly localized staff and facilities, plus the benefits of strong brands, customer relationships and an installed base. Huawei and Zhongxing, are by comparison, recent upstarts.

As Huawei vice-president Dr Zhijun Xu told me in Shenzhen in August 2000, Huawei didn't begin winning any significant business in China until it had won a number of export contracts. Said Xu, Chinese customers simply wouldn't take a risk on potentially inferior equipment until they had seen acceptance from their international peers.

Xu pointed out that quality, not price is the major determinant of gaining telecom equipment business in China. This makes sense. With massive growth rates, Chinese telcos have little reason to prioritize price when purchasing. Their biggest issue is making sure that equipment is scalable and durable enough to handle exponential increases in demand.

Xu also pointed out that Chinese companies no longer depend on foreign licensing and reverse engineering to develop their technical smarts. Companies such as Huawei are beginning to reap the benefits of the country's extensive investment in technical education and research. Unlike Western companies, many of those staff are also proficient in languages such as Russian, Arabic and Spanish, enabling Chinese companies to gain an export advantage in significant developing markets.

At least two Chinese innovations tend to bear this out. Two of the most exciting developments in wireless technology are the TD-SCDMA and LAS-CDMA platforms. Both are developed by Chinese companies, and if they perform to specification, will dramatically improve upon the core 3G CDMA platform developed by American CDMA specialist Qualcomm.

Indeed, foreign market shares in China have probably peaked. For companies such as Ericsson that currently earn billions in the country, the dramatic rise of Chinese-owned manufacturing capability is worrying. Huawei, Zhongxing and others such as Datang have made massive inroads into the Chinese market share of foreign vendors in just two years.

The illusion of the services sector

For others, the real prize is China's telecom services sector. As mentioned, Vodafone and News Corporation have already paid a hefty premium to buy small minority stakes in Chinese operators.

But these investments are highly risky.

Several years ago, some 46 Chinese and foreign investors placed a total of $1.4 billion investment in China Unicom, the rival carrier to China Mobile. Among these 46 investors were many of the world's top telcos.

Unicom was an initiative of a number of rival ministries to the telecommunications ministry and as such, never enjoyed regulatory equality to the incumbent.

In its core business—GSM cellular and paging—the company suffered discrimination. For example, it was handed a paltry 6 MHz spectrum allocation to run its national GSM network, when something above 15 was required. At first, it suffered difficulty even securing basic interconnect access to the national public network.

Worse for the foreign investors, their $1.4 billion investment didn't actually give them ownership rights over Unicom stock, for this was illegal under China's laws. Instead, it gave them access to a complicated "Chinese-Chinese-Foreign" investment scheme that promised a return of a mere 12%.

Under pressure from the incumbent, even this scheme was declared illegal. After much delay and acrimony, Unicom was forced to take out a loan to pay back the investors.

Since then, the law hasn't changed. Apart from Hong Kong issued stock, foreign companies still aren't allowed to own direct stakes in Chinese telcos. News Corporation's stake in Netcom is technically illegal, a fact noted with surprising candor in China's state-owned press when it was first reported in early 2001.

Although China's accession to the WTO would potentially change this illegal status, the fact remains that reactionary forces in the Chinese government could effectively outlaw News Corporation's holding at any time.

Buying what?

Vodafone's 2% investment in China Mobile, while legal, is equally risky. China Mobile is a Hong Kong-listed entity that is majority owned by the Ministry of Information Industry's incumbent mobile operator. This operator "sells" or transfers individual provincial networks to the HK entity in order to raise capital for its mainland expansion.

The latest transfer will see the HK entity pay $32 billion for seven provincial networks. Vodafone's $2.5 billion acquisition—which interestingly implies a total valuation for China's cellular sector of something akin to a quarter of trillion dollars—will be used to finance this transfer.

On the surface, Vodafone is buying into one of the world's hottest cellular markets—China has already overtaken Japan and is scheduled to overtake the US to become the world's largest in 2002. But closer investigation reveals that China's cellular glory days may be behind it.

For a start, monthly revenues per user are dropping by nearly 30% a year from the already low level of 178 yuan (about $21). This is already less than half the US level. Unicom's revenues per user are even lower, coming in at just 129 yuan (about $16).

These tariff declines are likely to accelerate as Unicom rolls out its network to more cities.

An early indication of this came when China Mobile offered a monthly tariff on its old analog network of just 30 yuan ($4) a month. Investors were so concerned that they slashed 17% in value off its HK stock price the week after the announcement.

Further evidence of the coming price war came when Unicom said that tariffs on its forthcoming CDMA network rollout would undercut GSM tariffs by as much as half.

Tariff cuts are necessary to increase penetration rates. China remains a poor country, with an effective annual income per head of US$3,300 (after adjustments are made for purchasing power parity). This is one-tenth the level of the US and one-sixth the level of neighboring Hong Kong.

With a penetration rate approaching 10% already, China's cellular operators have probably exhausted the possibilities for growth in that country's thin middle class.

To continue the sort of growth predicted—for example, Unicom expects to sign up 13 million customers on its CDMA network in the next year—investors will have to be prepared to suffer major cuts in earnings. With continued capital needs adding to earnings erosion, profits will be hard to come by.

Qualcomm's China hopes

In the light of this, it is amazing that a company such as Qualcomm has invested so much effort—and created so much public expectation—in attempting to crack the China market.

After years of delay, China Unicom announced in early 2001 that it would select multiple suppliers from a list of 12 to supply a US$1.8 billion national CDMA network in the country. Unicom was certainly ambitious. From a current network covering just five cities with half-a-million customers, it expected to add another 12.8 million in one year. Although it won't be selecting tenders until May, it hoped to get some of the new networks up and running by October 2001.

The problem was that Unicom had come this far previously. Some one year before, Unicom had likewise called tenders. But then in June 2000 came the announcement that it had withdrawn its plans. It wouldn't build CDMA networks until 2003. It would use local technology. At the time, analysts agreed it made little sense for Unicom to install a 2G CDMA network and that it would reduce risk if it waited a few years for 3G technology.

Qualcomm's Irwin Jacobs traveled to Beijing that September and managed to extract a turnaround from Premier Zhu Rongji and Information Industries Minister Wu Jichuan. Unicom would go ahead with the CDMA rollout, albeit at a delayed and reduced rate. In between, it wasn't clear how the accepted logic of the June decision had actually changed.

But Jacobs had to give away something to get the decision—according to a Beijing Dow Jones report of the time, Chinese manufacturers would only have to pay 1% on equipment sales and 2.65% on handsets. This compared with royalties of nearly 8% in other countries. On the basis of Unicom's announcements, Qualcomm would earn a paltry $20–80 million from network deployments over three years—chicken feed in the context of its $2.5 billion annual revenues.

Cheap handsets

It could do a little better on handsets but values are collapsing in that market. The successful entry of local Chinese handset manufacturers into the market is credited with dragging down wholesale prices by over 70% in three years. And although Qualcomm has entered into licensing agreements with these manufacturers, it remains to be seen how

Qualcomm will actually police and audit these companies to ensure that they pay their fair share of royalties.

Unicom's plans may suffer more setbacks.

The involvement of Zhu suggested that the CDMA plans were largely resurrected on the basis of China's original promises upon gaining accession to the WTO. US negotiators had made acceptance of Qualcomm's technology a key point of China's entry.

But one year after the US and Europe had agreed on terms for China's WTO entry, there will still issues. China was still refusing to accede to Western demands over agriculture, industrial subsidies and telecom entry. According to press reports, Europe and the US wanted commitments to be honored before WTO entry. China wanted WTO entry before it honored the commitments. Added to this impasse were political incidents such as the downing of a US spy plane on Hainan Island in April 2001. None of these events helped Qualcomm's cause—or the cause of foreign investors at large seeking access to China's telecom sector.

The illusion of Chinese telecom riches remained fool's gold. Not that this was an unusual case.

The Soeharto Seduction

Back in the mid-90s, when China's telecom sector was relatively moribund, the world's sexiest developing market was Indonesia. Under President Soeharto, Indonesia enjoyed political stability and was recording some of the fastest and most sustained economic growth rates ever seen.

In 1995, the Indonesia Government decided to accelerate teledensity growth by bringing in foreign investment under the so-called KSO scheme. This provided for joint Indonesian-foreign consortia to build and operate local loops across five separate areas. Government monopoly Indonesia Telkom would receive some 30% to 35% of the revenues and, 15 years later, would assume ownership of the infrastructure.

Foreign telcos were eager to gain the concessions, which they were forced to share with local Indonesian companies. Cable & Wireless, NTT, France Telecom, AT&T and Singapore Telecom were selected.

According to Asian business commentator Michael Backman, Australia's prime minister of the time, Paul Keating used his influence with Soeharto to get Telstra added to the winner's list.

Backman wrote of the incident, "Soeharto duly instructed the secretary general of the Telecommunications Ministry, Jonathon Parapak, to make sure that Telstra was included—and it was, much to the annoyance of NTT of Japan, whose share in the winning consortium was cut back to make way for Telstra."

Initial reports suggested that the KSOs were poorly managed. One Indonesian industry insider told me in 1997 that Telstra managers and NTT managers didn't consult each over procurement, leading to multiple orders for the same gear.

Indonesia Telkom, according to another observer, refused to give up its best technical staff to the KSOs, contributing to a chronic skills shortage. Local Indonesian partners added little value—one, Astra, was an automobile company! International telcos in the KSOs had as little as 10% to 20% of equity in these ventures, which of course, impacted on potential revenue and profit levels.

Rupiah crash

But the real problems started in mid 1997 when the Indonesian rupiah crashed, losing three-quarters of its value against the US dollar. The cost of imported equipment, almost always denominated in foreign currencies, soared. Tariffs, fixed at extremely low levels by the government, remained static. And as middle class incomes collapsed across Indonesia, so went any hope for the KSOs.

1997/8 was a disastrous period for both the Indonesian telecom sector and the economy. The economy lost nearly 20% of its value that year. And the KSOs found they couldn't afford to build—or find buyers—for the

0.66 million lines they were committed to building. In the end, they only reached 53% of their build target. The slowdown was even more dramatic when seen in the light of trade statistics—imports of telecom equipment to the country fell some 78% that year.

Over the next few years, the KSOs struggled on. In 1999, the KSOs increased total subscriber lines by just 3%. Although the government agreed to increase tariffs, high inflation and continuing depreciation eroded most of the gains. At the same time, wireless subscriptions surged doubling their number across the year to almost match the five year performance of the KSOs.

In 2000, the Indonesian Government bit the bullet. It announced a sweeping reform of the industry, announcing an end to Indonesia Telkom's monopoly and a broad liberalization schedule for the subsequent decade. Concurrently, Indonesia Telkom began negotiating to buy out the foreign partners in the KSOs. In some cases, partners were happy to get out and cut their losses. But in other cases, KSO investors resisted the offer. One KSO was offered a direct swap for Telkom's unwanted 22% stake in a minor Soeharto-family controlled service provider.

Telkom wins—again

As usual, the devil was in the detail. A proposed swap of assets between Telkom and international gateway monopoly Indosat in order to facilitate a duopoly clearly advantaged Telkom. The government agreed to a 45% hike in tariff rates over three years, but this was in the context of a 30% currency depreciation in the previous year and an inflation rate of 10%.

Throughout this all, Indonesia Telkom surged to unbelievable heights. As its KSO partners struggled to find any upside in the market, Indonesia Telkom announced that its 2000 EBITDA equated to a full 82.36% of revenues—possibly a world record for a telecom company. Even after all the standard accounting expenses were excluded, margins came in at a stunning 44.57%.

But foreign investors must have had their eyes closed to expect any better. Indonesia's telecom elite had always taken the long suffering Indonesian consumer and taxpayer for a ride. As Michael Backman reported in his Asian Eclipse book of 1999, in 1993, the state-owned Indonesian Palapa satellites were transferred to PT Satelindo, a company that was 60% owned by Bambang Trihatmodjo, son of President Soeharto. As Backman reports, "not only was there no tender, but it wasn't even clear that the government had received any payment for them."

Three years later, reported Backman, the Indonesian government called tenders for the privatization of a small stake in Telkom. As the conditions of the sale became clear, most bidders for the placement business pulled out for fear of violating the US' Foreign Corrupt Practices Act!

As the crisis started in 1997, Indonesia Telkom even awarded a contract to a company controlled by a Soeharto grandson, PT Arhista, which allowed it to keep 30% of all payphone revenues for simply procuring the payphones!

Thai twister
This type of dodgy behavior continues to this day.

In 2000, the ailing Thailand government—controlled by the Democrats party—decided that Thailand needed a third national cellular network, that would offer discount tariffs. A number of government agencies stepped up to the plate to sponsor the new company, named ACT Mobile.

But this decision had an interesting political complexion. The leader of the opposition Thai Rak Thais party was Thaksin Shinawatra. Thaksin happened to be Thailand's fourth richest man and had built up his fortune through a number of telecommunications businesses, including one of Thailand's two national GSM networks, Advanced Info Services.

ACT was to be the people's network, offering monthly fees of around 300 baht (US$6.50), a considerable discount to the 500 baht fees charged by AIS and its rival, Total Access Communications. Vendors, eager to line up at the ACT tender trough, eagerly contested the several hundred million dollars worth of contracts on offer.

Shinawatra's Thai Rak Thai party was swept into office in early 2001. But on the last day of the outgoing government's tenure in office, the soon-to-be ex-Communications Minister mischievously signed supply contracts for the ACT Mobile network with three vendors including Mitsui and Ericsson.

Shinawatra's new government was placed in a difficult position. Cancel the contract and critics would charge that Shinawatra was favoring his own business interests. But the new network was clearly against Thailand's broader economic interests. For a start, a new government-controlled network would have been a step-back for a country that wanted to liberalize and reform its telecom sector.

Additionally, the original ACT announcement had already achieved its intended effect, inducing both TAC and AIS to cut their tariffs to the 300 baht a month rate. Most importantly, Thailand's cellular penetration rate was just 6% with both networks experiencing light traffic loads. A third network would clearly be both a waste of precious investment resources and a threat to the overall economic sustainability of the cellular sector.

Shinawatra's government took the classic bureaucratic option—it announced a long-term review. As March passed, the first deadline for payment of vendors passed—without any payment showing up. The successful vendors publicly mused the option of legal action, but at least one was honest. One could sue the government, but you'd probably never do business in the country ever again.

Mahathir's folly

Foreign investors aren't always so easily taken in by the promise of telecommunications and technology markets in exotic countries.

A case in point is the Malaysian Multimedia Supercorridor.

Conceived in 1995 by Prime Minister Mahathir Mohammed, the MSC was Malaysia's grand attempt to emulate Silicon Valley. Some $5 billion was committed to develop a 750 square kilometer area stretching south of Kuala Lumpur into a hub for local and multinational IT&T

businesses. Multi-gigabit fiber links were laid. Special "cyber-laws", tax breaks and intellectual property protections were guaranteed. And three universities and a new administrative seat of government were established to help the area achieve critical economic mass.

Meanwhile, global IT companies were courted. A special advisory panel was set up, featuring the likes of Microsoft's Bill Gates, Sun Microsystems' Scott McNealy and even futurologist Alvin Toffler. Toffler raved about both the idea and Mahathir, describing him as the world's only Islamic leader who thought about the future and not the past. Microsoft committed to placing its SE Asian HQ there. The MSC idea seemed visionary and brilliant.

But then came the economic crash in 1997. Mahathir's worst populist instincts came to the fore. He blamed Jews and other foreigners for the country's capital flight and currency depreciation. Controls on capital movement were implemented. Mahathir's deputy prime minister was controversially arrested and imprisoned. Opposition newspapers and Internet sites were subject to official warnings and, in some cases, direct government pressure.

Foreign companies and the MSC's advisers voted with their feet. Toffler boycotted the panel. Only 38 "world-class" companies became involved with the MSC and few of them had committed substantial resources. By 1999, Microsoft employed just 15 employees at its MSC facility. The few who had taken advantage of the MSC's benefits complained of high property prices, government red tape and hassles in getting visas for foreign workers.

The writing had been on the wall from the start. Far from using the MSC to establish a free-wheeling enclave, the government had awarded Telekom Malaysia the exclusive right to provide telecommunications services there. Indeed, much of the MSC seemed geared towards creating nationalistic edifices. There was a new airport, situated some 70km from the city (subsequently, several significant air carriers discontinued service there).

Then there were the twin Petronas Towers—designed as tallest buildings in the world—situated symbolically at the northern-most point of the corridor. Absurdly for a cyber-city in a hot climate, entrance to these towers demanded a dress code—no jeans or shorts! To this day, vacancy rates in the towers remain high.

NTT & SingTel rebuffed

The biggest foreign investor in the MSC, Japan's dominant fixed line carrier NTT, soon found that Malaysia's newfound tolerance of foreign investment had limits. When it tried to negotiate a 20% stake in Telekom Malaysia, the deal broke down when it was told it would have no management influence.

Likewise, Singapore Telecom was rebuffed in an attempt to buy a small Malaysian carrier, Timedotcom, when Malaysia's political leadership provided support for a Malaysian rival. Foreign shareholders in Telekom Malaysia fled their investments, reducing their collective stake from 16% of the operator to 11%. MSC or no MSC, foreigners felt unenthused about the Malaysian economy and unwanted by the Malaysian government.

Six years on, the MSC has done little for Malaysia. Internet access rates are low by middle-income Asian standards—indeed, on a par with northern neighbor Thailand, which is a much poorer country. And while the MSC gleams with new infrastructure, it has attracted merely 9,000 new workers. Meanwhile, according to one Asian media report, nearly 500 schools elsewhere in the country don't even have electricity!

Foreign companies displayed an unusual degree of circumspection in minimizing their exposure to the MSC. Unusual in the sense that they have often been so naively gun-ho in other bubble markets such as China and Indonesia.

Chapter 8

Ask the user, stupid!

Alan Horsley is an unusual man. At first glance, he's a typical thick-set 50-something Aussie bloke. Ten years ago, he was buying the telecommunications for the state government of Victoria, Australia's smallest mainland state.

Then he packed in the security of his comfortable public service position and joined the Australian Telecommunications Users Group as its executive director. This group was no club for ham radio enthusiasts. It was a fully-fledged lobby group, supported by membership fees and financial support from hundreds of Australian businesses and organizations. It was sufficiently influential that it was automatically granted a place at the table of every official or semi-official government inquiry into telecommunications.

Then, five years on, Horsley received the call from the government. They wanted him to sit on the board of the country's dedicated telecom regulator, the Australian Communications Authority. What makes Horsley so unusual? He's not a career policy bureaucrat or an ex-monopoly operator executive. Indeed, he is the first representative of telecom user interests in a position to actually administer and execute government telecom policy.

Horsley's appointment was an indication of the esteem in which he and his group was held. Since its formation in the late 1980s, ATUG had consistently called for competition and a better deal for Australia's corporate users. In 1997, it got it when the market was completely liberalized. The year before, the Liberal Party's election policy on telecom read like a direct lift from ATUG's wishlist. When it was elected to government that year, ATUG's members were elated.

That said, Australia remains an anomaly. It is the only country in the world where its dedicated telecommunications legislation enshrines "long-term user interests" as a key objective of policy.

An organization such as ATUG is most effective in a pluralistic, non-corporatist environment. ATUG's original backers came from across the spectrum, ranging from Australian Associated Press (which operated an extensive private network), through to the Australian Navy.

Influential user groups have emerged elsewhere. One of the more notable is the Hong Kong Telecommunications Users Group. It can trace its origins to the Hong Kong Jockey Club, which maintains a lucrative betting monopoly in the city. With an extensive branch network and phone betting operations, the Jockey Club is one of Hong Kong's biggest telecom users. Under colonial rule, it was also one of the more powerful private institutions in Hong Kong.

HKTUG isn't as powerful as ATUG but it plays an important role in informing Hong Kong's telecommunications debate. Hong Kong's independent telecom regulator, the Office of the Telecommunications Authority, is probably the most transparent such body on the Asian mainland. It consults with industry to an extraordinary degree when compared with its peers.

Breaking down Europe's walls

Europe's International Telecommuncations Users Group, which boasts financial support from the likes of American Express and Shell, is similarly influential. INTUG's formation was actually encouraged by the European Community policy bureaucracy when it became concerned that it was overly informed by pro-monopoly views from the continent's national telcos and the organized labor that worked for them.

To this day, many telcos regard user groups with suspicion. The feeling is mutual.

INTUG chairperson Diana Sharpe says that telcos still don't really understand how to cater for multinational corporate users and that they still don't provide genuinely competitive offerings.

In a speech delivered in South Africa in 2000, she said "At present there is very limited real choice for users. What business wants is effective choice and reasonable prices. Today, it can only be found only in the centres of a few major cities and on the routes between them. It is fine in the 16th Arrondisement of Paris where the OECD resides, but very hard to find at home in rural Suffolk".

"Businesses are also looking for partnerships with operators. However, the operators seem to be struggling to understand this. They prefer to offer something more rigid."

Danish telecom consultant Allan Fischer-Madsen, who heads INTUG in Europe, agrees. "It is only the beginning of the end of the tyranny of distance. Its death has been much exaggerated. The vast majority of tariffs still depend on distance and where they cross national borders the prices can jump out of all proportion," he said.

"It is possible to find high levels of competition in London and Frankfurt-am-Main and between the two cities. It is a different matter beyond the suburbs, in smaller towns and the countryside. Equally, there are many segments of the market, such as large corporations, SMEs, residential and so on, not all of which have yet seen the benefits of competition."

Multinational musical chairs

Certainly, the major established telcos are confused as to how to address the multinational user market.

For the past eight years, these telcos have gone through a confusing array of marriages and divorces. This has resulted in losses of hundred of millions of dollars.

The contortions of AT&T and BT have been the most painful. BT started off its global strategy in the mid-nineties by forming the Concert alliance with MCI, then primarily a US long distance carrier.

Against them was WorldPartners, headed up by AT&T but which included significant numbers of national telcos from around the world, particularly Asia.

France Telecom, Deutsche Telekom and Sprint formed their own Global One joint venture in response.

The mission of these companies was to provide seamless global services for multinationals. Services such as one-currency billing and multi-lingual customer support were promised.

But one by one, each alliance fell apart.

Concert came under pressure when MCI merged with Worldcom, forcing BT to look for a new partner.

Breaking up WorldPartners

Around the same time, AT&T realized that while many of WorldPartners' customers were US multinationals, it was the foreign partners in the alliance who were extracting the lion's share of revenue from the business.

As the company's president of the time, Simon Krieger admitted to this writer in June 1998, most multinationals are conventional customers in the sense that they mainly make local transactions in the countries in which they operate. These transactions were bound up in the revenues of AT&T's partners, and thus, weren't even recognized as WorldPartners' sales.

When Michael Armstrong ascended to the top of AT&T in 1999, one of his first decisions was to make an overture to BT for a JV. Ian Vallance, the now-resigned chairman of BT, happily reciprocated. That said, BT clearly saw AT&T as its second choice. As one senior BT executive told me at the time, "MCI was still our preferred choice of partner. It was a combination of the world's best market defender with the world's best market attacker."

The new BT-AT&T combination, which took over one year to decide on using the recycled Concert name, was more of a genuine joint venture than an alliance. Infrastructure and all global traffic were rolled into the venture. Even so, it struggled to make decent profits, apparently just managing to break-even in late 2000. With both AT&T's and BT's senior management facing unprecedented shareholder criticism in early 2001, it was unclear how long Concert might last in its current form.

But while AT&T and BT were muddling along, Global One was racking up losses—and massive ones at that. In 1999, it lost over $600 million, some five years into its operation!

The joint venture wasn't quite sure what it wanted to be—like WorldPartners and Concert, it targeted multinationals, but it also offered calling cards, residential services in some countries and even bid for infrastructure projects in such unlikely places as Hanoi in Vietnam and rural China.

Global One also offered dreadful service. It admitted, for example, that in February 1998, its average European data customer suffered nearly nine hours of downtime for the month.

So at the end of 1999, Deutsche Telekom and Sprint simply pulled the plug, leaving France Telecom to save the sinking ship. With the loss of the two shareholders, went some of its key country managers—creating further challenges for the group.

Cable & Wireless was a late starter in the global multinational market, previously preferring to keep its global subsidiaries at arm's length from each other. But after selling off profit makers Hongkong Telecom and Optus in order to concentrate on the global corporates, the strategy wasn't looking so smart.

After just nine months of operation, the company announced a revamp of its global unit in early 2001, accompanied by 4,000 job cuts. With the company making losses on its global IP network, its sole remaining profit earner was its collection of small island monopolies in the Caribbean and the Pacific Ocean. Even that easy cash source suffered a limited future—five

Caribbean nations banded together to end the British company's monopoly in the region.

So who gets it right?

There's one thing in common between these dysfunctional global telcos. They all have monopoly backgrounds with a specialty in consumer markets.

Between them they have demonstrated little aptitude for profitably addressing global corporate markets.

Indeed, the most successful service providers in the multinational market have an altogether different attribute. They were once customers themselves.

The best example is Equant. This Amsterdam-based operator had its origins in 1949 when the world's airlines got together to form a harmonized communications network. This rapidly became the most extensive private network in the world extending to virtually all places where commercial jetliners travel.

Then in 1991, this network was spun into a new company named Scitor, with an eye to offering services to customers in other global vertical markets. It believed it understood the demands of these customers better than the generally clueless national telcos of the time. The subsequent expansion proved successful, garnering a whole range of new customers from the transportation, mining and hospitality industries. By 1997, the company was so successful that it was floated and re-named Equant.

Subsequently, the company has continued its expansion, extending its vertical solutions to encompass government, manufacturing and financial services. Its most recent customer wins include Coca Cola, Mitsubishi and Wrigley. In 2000, it increased its revenues by 43% to nearly $1.5 billion and recorded a small loss, reflecting some one-off acquisitions and alliances. Underlying EBITDA increased 13% to $180 million.

Specializing in financials

Another successful operator that grew out of a private network was Saturn Global Network. This company was founded in 1995 by M.W.Marshall, an UK money broker that had constructed a private network spanning several continents. Saturn became successful in the Asian region, where its vertical solutions for the financial services industry proved popular. In 1999 it merged with IXNet, an UK operator with the same financial services focus. The combined operation was a powerhouse, claiming nearly 2,000 financial industry customers worldwide, including some 60% of potential customers in the City of London.

As the company's Asia-Pacific managing director Drew Kelton told me that year, IXNet provided a full extranet solution for its customers, ranging from voice tie-lines to desktop financial information displays.

With its grounding in the financial industry, IXNet didn't attempt to force trendy but potentially sub-optimal technology solutions on its customers. No one was forced to use VoIP instead of carrier-class switched circuits. As Kelton pointed out, any loss of quality on a voice tie-line could result in a garbled order and potential mistakes worth millions of dollars!

In late 2000, both IXNet and Equant succumbed to the forces of consolidation and were snapped up by larger suitors. IXNet became a division of Global Crossing while Equant "merged" with France Telecom's wholly owned Global One operation.

In both cases, the acquirers claimed that the buy-outs would enable operating efficiencies and reduced costs. This may be true. Both IXNet and Equant ran expensive and thin circuits across the world that will benefit from aggregation.

But with the buy-outs comes the risk that they will both lose the company culture that made them attractive to their customers. There is no doubt that their respective origins as vertical private networks endowed them with unique insights and a pro-customer philosophy. Whether France Telecom and Global Crossing can preserve these advantages remains to be seen.

Residential dysfunction

If the story of multinational corporate telecommunications has been one of dysfunction and missed opportunities, then the same could also be said for residential telecommunications.

There is no doubt that competition has proved effective in cutting tariffs to end users. One of the more ubiquitous sayings in the US telecom industry is that long distance is 5c a minute away from becoming free. As anyone who has churned in the US knows, the ensuing onslaught of night-time telemarketing calls from long distance operators is testament to the aggressive competition in the sector.

But despite regulations enabling the unbundled local loop and access for CLECs, few US residential customers can yet select an alternative provider for their local service. Quite simply, the regulated price caps for local services prevent competition and entrench existing monopolies. Cable telephony services have yet to take off either. One report in late 2000 suggested they had attracted less than half a million users across the US.

Even the benefits of competitive long distance telephony have been mixed for many customers. Some US long distance operators have made criminality their specialty, engaging in all types of dubious activities. Slamming—the act of signing up a customer without their consent—and, cramming—the act of billing them for calls they did not make—are two of the more common violations.

The pre-eminent historian of US telecom misbehavior is William Van Hefner, a long distance consultant based in Northern California. Over the years, Van Hefner has reported and archived information in his Hall of Shame at *www.thedigest.com*

In one case reported by Van Hefner, a company called Telecom Services Corporation illegally slammed 70% of the residents of Colville, Washington to its services. In another case, staff from a Californian company called Cherry Communications impersonated Pacific Bell employees in order to gain access to customers' homes and change their lines.

Another Californian company, WorldxChange, had its operating license in that state revoked for three years after it slammed 57,000 people—mainly Asians and Latinos. The company ultimately paid $21.6 million in settlements and fines. Van Hefner revealed that the company's principal, Roger Abbott, had previously been convicted for cocaine dealing and had run a precious metal investment scheme which caused some to lose as much as $250,000 each.

The real Ronnie Biggs of US telecom, however, is Daniel Fletcher. Van Hefner says Fletcher's companies illegally slammed hundreds of thousands of customers. He is now a federal fugitive, having apparently fled the country owing $8.7 million in unpaid fines and carrier payments.

The wide extent of criminality in the US competitive telecom sector has been noticed elsewhere. Australia, for example, has implemented a Telecommunications Industry Ombudsman to deal with customer complaints about telcos and ISPs. This Ombudsman has the power to make determinations of up to $US25,000 in favor of customers—directly invoiced to the offending carriers. Currently, it is receiving about 12,000 complaints a month.

The biggest emerging scam in Australia and elsewhere is so-called Internet dumping. This is where an Internet user is unknowingly dumped from their dial-up connection and reconnected to a premium rate or IDD line.

Network unreliability

Corporate users haven't suffered as much from the effects of outright fraud. But they usually pay more for the same service than residential customers—in many markets, business charges are often twice those of residential charges for exactly the same product.

Survey after survey has shown that business users don't mind this so much as they are not especially price-sensitive. But the implied part of this deal is that for a higher price they receive better quality-of-service and reliability.

Unfortunately, too often, this has proven not to be the case. And the unprecedented round of mergers and acquisitions didn't help.

One notable example came in the second week of August,1999. A botched software upgrade forced MCI Worldcom's United States' frame relay network out of action for up to ten days, disrupting service to 3,400 customers. Unfortunately for MCI Worldcom, one of these customers was the world's largest futures and commodity exchange, The Chicago Board of Trade, which lost no time in going public with its gripes—to the extent of threatening lawsuits. Central to the complaints—MCI Worldcom executives appeared to have no idea what was wrong with the network and did not provide adequate information updates.

It later emerged that the old MCI frame relay network operated fine throughout the ten-day outage and that the problems could be attributed to the Worldcom frame relay network. The two had never been successfully integrated.

Cable cuts

Rushed installations of fiber optic cables are also proving problematic. The US-China undersea cable was cut on two occasions in early 2001. Although carriers were quick to blame lax regulation of fishing boats for the cuts, at least one consultant close to the industry blamed the industry.

Said UK-based Brian Powell after the second cut, "This just goes to prove the stupidity of launching a major subsea network without having full redundancy in place. This is typical of American mentality to be first to market and start making a few bucks at any cost."

"Well it certainly has cost them—what this has cost them is the reputation of the China-US cable—and as soon as other new trans-Pacific cables come on stream later in 2001 & early 2002—just watch all these ISPs and IP backbone providers that have been burned twice in less than two months desert China-US like a swarm of flies."

"This second catastrophic failure in less than two months brings seriously into question the robustness of the system design, the level and quality of engineering that went into the construction and the corners that have been cut in route planning to save a few bucks. Not only is this area heavily fished, it is also a major shipping channel like the English Channel. Now you can see why no Transatlantic cables are brought up through the English Channel too risky—they all land in Cornwall, Devon or Wales."

"The sea depth for the entire route of the cable is very shallow and to make matters worse, Chongming is an island in the Yangtze estuary opposite to Shanghai—the whole vicinity is anchorage for Shanghai Harbour."

"Because this cable is laid so close in to the shoreline in a heavily trafficked & fished area, it should be buried at least 1m under the sea bed in steel duct. I think a more robust solution would be to make this section terrestrial (like FLAG is across Egypt & the Thai peninsula)."

"It will now be very hard to sell capacity on this cable and nigh impossible once newer trans-Pacific cables come on stream—even when the redundant link is working—the reputation has been burned already. After all—who is going to fly an airline that had two 747 crashes in as many months?"

Chapter 9

The collapse of value—and promise

Like the thylacine and the plesiosaur, the bandwidth bubble isn't necessarily extinct. Indeed, there are still regular—if surreal—sightings.

A case in point was a conference on optical networking held in Anaheim, California in late March 2001. The OFC show, organized by a pre-eminent American engineering association, attracted nearly 40,000 visitors, apparently twice the previous year's attendance.

Hundreds of exhibitors demonstrated equipment for which outlandish claims were made. The sales reps for the magazine I worked on at the time were excited. The telecom boom wasn't over yet!

The first all-optical switch. Lasers without external pumps. Thousands of wavelengths at less than $100 each. New heights in amplification effectiveness. An engineer's wet dream.

To cap things off, George Gilder made an appearance. As attendees polished off their fruit tarts, Gilder provided the after-dinner entertainment. Later he was to show that he was suitably impressed by proceedings. "While the market still is saying no to the new network topology, OFC issued a resounding yes", he wrote in his subsequent newsletter.

But there was one problem with all of this. Like the glittering but fake fantasies at the adjoining Disneyland, the technological breakthroughs of OFC were a mirage.

Many of OFC's exhibitors were the last cabs off the bandwidth bubble rank. Even as stock prices plummeted throughout 2000, venture capital firms continued to plough funds into optical startups. There was no shortage of willing innovators. A good idea might be quickly snapped up by a JDS Uniphase, a Nortel or a Cisco. By early 2001, some 70 optical and switching minnows had been acquired by a handful of industry giants.

The buzz at OFC was the dying vestige of this phenomenon.

Not that optical networking was sinking.

That's fairly hard for a technology barely out of the laboratory. But one look at all the prospective customers for optical switches—the giant struggling incumbents, the debt-laden fiber barons, the cash-strapped CLECs—and it was difficult to see how all but the smallest number of OFC exhibitors would actually make their mark in the real world.

But if the all-optical switch had a bit of the Mickey Mouse about it, that wasn't yet apparent to everyone, despite the disappointments of the previous year.

Faith in elasticity

A few weeks after OFC, an international edition of the Financial Times nonchalantly asserted that a planned 8 terabit cable connecting Singapore and an Indian city would be "utilized almost immediately."

Right? The same week, researcher Telegeography issued a fascinating report.

"One demand model estimated that trans-Atlantic Internet capacity would equal 175 Gbps in 2000, but TeleGeography's survey data from that year found that aggregated Internet backbone capacity only reached a third of that number."

"Moreover, TeleGeography's inventory of "used" capacity seems small when compared to the vast supply of lit capacity. In 2002, TeleGeography projects that the supply of trans-Atlantic bandwidth will total 3,494 Gbps.

Even if Internet backbones quadruple in size, and other networks experience double-digit growth, bandwidth deployed by carriers and ISPs will amount to only 532 Gbps."

"In other words, used bandwidth will account for less than 20 percent of estimated supply." The same week produced more incongruous news.

Pyramid, not renown as a particularly negative or contrarian analyst firm, issued a fascinating report on the arcane science of price elasticity. In the face of declining prices, extremely positive elasticity is the only way that the telecom industry can work off its excessive inventory of bandwidth.

Pyramid's report claimed that the international voice market had been experiencing negative elasticity for two years—in other words, volumes weren't increasing enough to balance the effects of price drops.

When considering the double-digit revenue losses experienced by the IDD divisions of the likes of AT&T, Telstra, Singapore Telecom and Hongkong Telecom over that period, this seemed like a reasonable observation.

But Pyramid then went on to claim that the corporate data market would experience negative elasticity in 2002. And what's more, the Internet would experience the same effect in 2005. In other words, the Internet would experience only four years more growth before it entered a financial decline! Such gloomy figures were borne out by additional data. US telecom capital expenditures grew by 32% in 1999 and 25% in 2000, according to Morgan Stanley.

In 2001, expenditures would grow by just 2% to 3% and would fall by the same amount in 2002. Merill Lynch was more negative, predicting a 10% drop in spending in 2001 and a further 5% drop in 2002. Another study from Lehman Brothers, suggested that emerging US carriers would incur a collective financial loss of $10 billion in 2002.

The incredible shrinking China

Even that great sunrise market, China, looked gloomy. If the view of some Hong Kong stock analysts was any guide, the China telecom equipment market was undergoing some negative elasticity of its own.

Price competition for the business of China Telecom and China Mobile was apparently so intense that they undershot their capital expenditure budgets by around 30% in 2000. Future savings of 15% annually were anticipated.

China wasn't the only national market throwing up concerns.

In 1999, the Australian Government commissioned a national bandwidth study from a number of private consultants. Many expected the report to say the usual things—there needs to be more government promotion of the benefits of competition, there is a need for more and cheaper bandwidth, etcetera etcetera.

So it was quite a surprise when the inquiry found that only between 0.15% and 1.69% of the country's planned interstate capacity (depending on the route) was actually utilized, while on the country's allegedly scarce international conduits, only 21% of capacity was used.

Despite this abundance, the Australian attitude to bandwidth is stingy. This is largely because of continuing high prices and the realization by many of Australia's Internet service providers that they don't need ever-increasing international links.

As reported earlier in this book, Telstra actually reduced its US Internet backbone bandwidth last year. Indeed, Telstra has regularly displayed an aversion to paying for Internet bandwidth to the US, claiming that it subsidizes free-loading US users to the tune of some US$250 million a year. The solution? More hosting and peering infrastructure. Australia's 910 ISPs, many of which are fairly marginal enterprises, were similarly keen to reduce their international connectivity costs.

Bandwidth demand growth is also looking less than exponential elsewhere. One report from Jupiter MMX forecast that European broadband penetration would only reach 14% by 2005.

This isn't necessarily because of consumer indifference towards broadband. For example, German ISDN penetration equals 25% of fixed line penetration. But both the German ISDN provider and market seem to be both content—and inert—for the moment.

Likewise, broadband has been a relative non-event in Asia. There are less than 100,000 broadband users in China and probably little more in the rest of south-east Asia combined.

The major exception is South Korea, the only market in the world to boast of higher broadband penetration than the United States.

Around 4 million households—or half the total—subscribe to a DSL or cable modem service. Key to the growth has been intense competition and cheap prices of around US$30 per month.

Already the industry has seen a few casualties and some analysts wonder if competitors to Korea Telecom such as Hanaro Telecom can last the distance.

The major drivers of broadband usage in Korea are streaming audio & video, and interactive gaming. High usage of these applications has yet to be observed in comparable Asian markets such as Taiwan and Hong Kong.

No hope for voice

If data and Internet traffic won't quickly fill the pipes, then what about voice? Here the prognosis also looks gloomy.

International voice was traditionally the most profitable part of the telecom business. For years, it drove billions of dollars worth of profits for developing world telcos under an arcane international accounting system. Because rich countries originated more calls than poor countries, the settlement system created a massive transfer of funds from Western telcos to the developing world.

At one stage, the ITU estimated that this transfer probably equaled some $8 billion annually. Countries with large populations such as China, India, Mexico and Indonesia were the greatest beneficiaries. In his 2000 book "You Say You Want A Revolution", former FCC chairman Reed Hundt claimed that America's outbound payments "exceeded the total of United States' non-military foreign aid."

America's relatively inexpensive rates exacerbated its telecom balance-of-payment problems. A cheap but clever by-pass system named "call-back" enabled operators to offer cheap calls to customers in developing countries by arbitraging American tariffs. Although these services were offered illegally in many countries, they quickly became popular with price-sensitive IDD callers worldwide.

In 1996 and 1997, the FCC struck out. Buoyed by a WTO agreement that would see most of the world's major telecom markets opened to competition and foreign investment, the FCC decreed a most extraordinary requirement on US carriers.

It would now be illegal for them to pay more than a FCC-prescribed price for international connectivity. Carriers from developing countries were predictably outraged that a US government agency was naming the price they could charge for access to their sovereign networks.

They had a point. The FCC's new pricing regime wasn't based on network cost information. Instead, it divided the world into three tiers—high income, middle income and low income—and decreed arbitrary prices and phase-in dates for the new arrangements.

For rich countries, the suggested tariff of 15c per minute wasn't unrealistic, as many had already introduced competition. But for poor countries, the FCC's mandated tariff of 22c was 60% to 80% below their existing tariffs. The ITU was broadly supportive of the FCC's cause, but offered its own preferred reform program, which entailed shallower and slower price cuts.

These changes were no small matter for government-owned operators—they typically used margins on IDD to cross-subsidize universal service. They also had—and continue to suffer from—limited access to alternate sources of financing.

Nevertheless, the US telcos enjoyed considerable success in gaining cheaper rates on foreign networks. AT&T's Singapore-based vice president Phil Overmyer told me in 1999 that AT&T had managed to reduce rates to almost all Asian territories except Diego Garcia, the UK-territory in the Indian Ocean where Cable & Wireless enjoyed a monopoly!

At the same time, Asian executives such as Singapore Telecom CEO Lee Hsien Yang and Hongkong Telecom CEO Linus Cheung complained of the heavy-handed attitude of the FCC—an attitude that certainly didn't help the American operators in their general relationships with foreign operators.

The collapse of IDD pricing

The US-led push for cheaper IDD pricing has largely succeeded. In the mid-90s, US call charges to foreign destinations averaged as much as $1 per minute. Best prices have now fallen below 25c a minute for major destinations in Western Europe and Asia. Other countries with vigorous competitive regimes have enjoyed similar benefits. Australia's Cable & Wireless Optus offers off-peak calls to destinations such as the US for as low as US2c per minute.

For the many developing countries that have retained their monopolies, the pain continues.

The successor to call-back is IP telephony (not to be confused with Internet telephony). Unlike a conventional phone call that occupies an entire circuit for the duration of the call, IP telephony converts the transmission into IP packets. This increases the carrying capacity of networks by up to ten times, allowing cost-efficiencies of as many as six times over conventional networks.

IP telephony can be used to secret calls into countries that maintain monopolies. Typically, this is achieved through running the calls through leased lines operated by legitimate Internet service providers. Some countries have arrested ISP owners for precisely this type of activity.

Other developing countries have chosen to embrace, rather than fight, IP telephony. In China, it is offered by the incumbent, China Telecom, and a number of competitive licensees—as a cheap alternative to regular IDD telephony.

The service has proved extremely popular and provides China's government with an effective way to test a provisional competitive regime without risky industry restructuring. IP telephony has also played a key role in facilitating competition in Hungary.

IP telephony is of most benefit to competitive upstarts with packet network investments. Although traditional carriers could potentially use IP telephony to buttress their declining IDD and long distance margins, in reality they have expensive investments in existing circuit-switched networks.

Most affected by the rise of packet networks were the competitive long distance carriers such as Telegroup and Pacific Gateway Exchange that emerged in the mid-90s. As they achieved critical mass they made extensive investments in Class 5 switches across multiple international markets.

The rise of IP telephony and the fall in tariffs has all but wiped out their prospects.

Declining fortunes

The ITU reports that international circuit costs fell by as much as 72% in 1999. The trend shows no sign of slowing in the light of the growth of packet-switched networks.

In March 2001, the ITU convened a special global forum on IP telephony.

Despite opposition from some African and Middle Eastern countries, the final ITU view was that IP telephony was a positive development that should be encouraged.

Ominously, but realistically, it said that while there would be losers from IP telephony, the overall public good would be better served from the lower prices and innovative services that would be created by the platform.

By early 2001, some 6% of global IDD traffic was carried over IP networks. Competition was spreading to as many as 70 countries by the end of the same year. Voice telephony, reduced to cost on the international high roads and usually regulated to cost in local markets, was well on the way to becoming as much of a commodity as power and water. The major difference? It was more abundant than either of those resources.

Chapter 10

The long, dark years

By the end of the first quarter of 2001, it was clear to almost everyone that the global telecom industry was in crisis.

In a twelve week period over February, March and April, telecom manufacturers announced plans to eliminate 130,000 jobs. This was equivalent to the entire communications workforce of Australia and New Zealand.

Cisco headed the list with planned cuts equivalent to 18% of its workforce. Lucent was cutting 12% off its numbers, Ericsson 11% and Motorola 8%. Not one major vendor was immune from conditions. Ascendant vendors such as Marconi and Sycamore—that were claiming market share from rivals—were also cutting numbers.

Even Nokia, which had been increasing its share as the world's leading cellular handset manufacturer, was warning that global handset shipments in 2001 were likely to fall tens of millions of units short of expectations.

The sheer scale of the downturn defied belief. Barron's reported on April 21 that telecom companies had defaulted on US$6 billion of debt in the first quarter of 2001—some $4 billion of that was in March alone.

But not everyone was convinced. In that same issue of Barron's, a report by Bill Alpert attempted to debunk the notion of a bandwidth glut. His argument, as borrowed from Level 3? To that date, there were 41 cross-country fibers deployed in the United States with another 570 announced for deployment by 2003.

This sounded scary, but by Alpert's assertions, at current rates of capital expenditure, those fibers wouldn't be lit up for another 33 years. Hence, there was less bandwidth coming on line than expected and prices would not fall as fast as feared. Unfortunately, Mr Alpert missed the point. Why were telecom companies planning to spend money now on installations that wouldn't be lit until 2034?

What was the justification for making investors pay now for assets which wouldn't earn a return for nearly four decades? What investor alive now would really want to buy into such a business model?

The immediate future

Bad news on this scale is like a red flag to some contrarian investors, especially those who still subscribe to the promise of a giant broadband market. Throughout March and April 2001, NASDAQ technology stocks would slide, only to rally with daily increases of as much as 10%.

Was this volatility a mere result of the phenomenon of the "dead cat bounce" or something more meaningful? After all the viciousness of the 2000 stock market correction and the wide spread staff cuts across the industry could constitute an effective "capitalistic cleansing"—clearing the decks for a quicker recovery and a return to the normality of double-digit sectoral growth.

But many of the industry issues identified in this book will not disappear overnight. For example:

Technology cycles for major deployments are lengthy and risky. The examples of Iridium and Globalstar aren't unique to satellite networks. Innovative disruptions in the field of optical switching introduce a new element of risk to long-haul fiber optic deployments. Given that the cost of optical equipment equals the actual fiber deployment, incorrect technology decisions could prove fatal.

IP telephony will promote competitive arbitrage—and reduce sustainability
The foundering prospects of American competitive operators will be seen elsewhere. IP telephony is proving successful in quasi-monopoly markets such as China and Hungary, but one could argue that it more an arbitrage device than a legitimate upgrade path for existing operators. Many developing markets may suffer overbuild of circuit-switched and IP networks as a result, resulting in the same sort of metropolitan capacity glut now seen on backbone routes. Many of the pioneering IP telephony operators won't survive—look at the poor survival rates for callback and independent switched long distance operators.

Governments will continue to intervene in deployments, and more so. As broadband access becomes more prevalent, so will the pressure on governments to do something about the so-called digital divide. This has already been observed in countries such as Australia, where the government there has dedicated hundreds of millions of dollars in privatization proceeds to telecom and Internet projects in rural and remote areas. In the future, watch for local and provincial governments to become actively involved in infrastructure deployment. Already, city authorities in places such as Chicago, Toronto and Canberra are either considering or deploying city-owned fiber networks, despite a lack of evidence of market failure.

Many of the disruptions to markets will come through opportunistic arbitrage This won't just apply to long distance, but also to wireless. A case in point is US trunked radio operator Nextel, which used its cheap 500 MHz spectrum allocations and a cleverly-adapted technology to launch a successful entry into the cellular market. Watch for Sprint and Worldcom, who control spectrum for fixed wireless using MMDS technology to try and convert their licenses to 3G mobile use. This is entirely legal under current FCC regulations.

Break-out successes such as Korean broadband and Japanese 3G will encourage others Bad economic conditions won't discourage operators from attempting to imitate proven successes across the world. Operators in Asia will expect that they can emulate the success of Korean cable and DSL providers in signing up one in every two Internet households to broadband, just as many operators believe they can emulate the success of NTT DoCoMo in signing up one in every five Japanese to I-Mode, a proto-wireless Internet service. Whether these other operators can do so at a profit remains to be seen.

Internet market growth will continue to slow From a doubling every 90 days to a doubling every 365 days, expect the deceleration of the Internet to continue. By April 2001, Level 3 was suddenly talking of a doubling every 18 months! But it's the state of what's on the Internet that could prove problematic for its future prospects. One of the biggest drivers of broadband demand was peer-to-peer MP3 file sharing. This is now being effectively litigated, regulated and priced out of mainstream acceptance. As dot-com content providers run out of cash, high-quality free content will reduce in scale. This will serve to reduce consumer interest in the Internet. There is good news, however: this effect should be partially offset by the development of non-English language content and resulting growth in global user numbers.

Brands and owners will win If you don't own essential access infrastructure or a good consumer brand, you won't have a future in telecom. While the advantages of the former are apparent, the value of a strong brand has been under-rated by many industry analysts. It will be interesting to watch the fate of Richard Branson's Virgin Mobile brand in the UK, Singapore and Australia. If his resale operations succeed in these markets, others with strong horizontally-integrated brands may be encouraged to enter the market. The future of bundles may not lie with telecom services as such, but instead, a more varied package comprising of consumer, retail, financial and communications services.

Diversification will triumph If brands count, then so does diversification. Pure-play telecom and Internet companies are toxic in the eyes of financiers right now. Infrastructure companies such as cable TV operators, energy utilities and conglomerates such as Hutchison Whampoa enjoy solid cashflows and a better reputation in the eyes of finance markets, and are the likely new source of telecom competitive activity in years to come. Look for existing telecom companies to engage in innovative strategies to restore their luster with financiers—for example, by prioritizing their unfashionable but cash-rich directories businesses, or even by seeking to gain revenues from non-telecom sources such as finance services. Level 3 props up its revenues with three American coal mines—while I doubt many will follow its example, it's a better strategy than totally relying on the stock and venture capital market for cash.

The promise of developing markets will remain just that As the Indonesian, Chinese and Thai examples in this book demonstrate, emerging markets are fraught with hazard for the unwary. Over the past few years, billions of dollars have been poured into markets such as these along with Russia and Brazil, both in the form of vendor financing and direct operator investment. As the political disputes with countries such as Japan, China and Mexico have shown, a signed WTO telecom treaty offers little more than an entry into a second round of the fight for market access. It will probably take decades for China's intellectual property and investment laws to catch up with the reality of the WTO ideal and by the time it does, local companies will be internationally competitive.

Telecom companies will now need business cases One of the underlying causes of the bandwidth bubble was the intense competition between vendors. All too often, telecom companies used vendor financing to fund their rollouts—clearly because those rollouts would not stand-up to the scrutiny of independent financing. Another practice coming to and is the gift of vendor stock to chief technology officers who purchased equipment from that same vendor. Such practices clearly undermined the process of rational investments.

Technology disruptions will continue to undermine industry sustainability With the hundreds of billions committed to 3G CDMA deployments, operators would at least expect to get a ten to fifteen year life from their investments. But a growing threat to CDMA comes from OFDM, a potentially superior technology that some are all ready dubbing 4G. OFDM could be safely dismissed as vaporware, but for the reality that many vendors are keen to avoid the economic rents that accrue to Qualcomm from CDMA. The technology choices available to today's operators—DSL, cable modems, IP, ATM, free-space optics, microwave, gigabit Ethernet, ISDN, CDMA, OFDM, photonics—are rich and exciting, but hardly predictable if one wants to still be in business some five to ten years out from now.

Much of the hype and fervor that accompanied the bandwidth bubble was the result of the assumption that basic and enhanced telecom access was a human want, desire and right on some sort of Maslow scale with food, shelter and love.

The developed world needed broadband Internet and needed it now. The developing world needed a telephone in every home and an Internet connection in every school and office. It was such a basic assumption that policy and strategy—whether it be in the Wall St boardroom, the FCC HQ in Washington DC, the Zhongnanhai compound in Beijing or the presidential palace in Jakarta—was universally informed by the same desire to achieve economic and social salvation via connectivity.

While it would be churlish to suggest that telecom access isn't an important or desirable goal, it is also important to remove industry-centric blinkers when considering optimum policy and strategic responses.

The household trap

At least one developing world telecom official, from Malaysia, has complained of the Eurocentric ITU's method of measuring international teledensities—using telephones per person. As he pointed out, household sizes are larger in the developing world than the developed world. Simple comparisons between the two are less than meaningful.

Indeed, when considering telecom markets in developing countries one should look for measures of public access, rather than private subscribership, when assessing market realities.

As a case in point, I'll use the example of Phimai, a 50,000 population town in north-eastern Thailand. Much of the population in this town live in villages clustered around the urban center. In these villages, the most common form of telecommunications is the cellular phone, although admittedly ownership is the preserve of family heads and village officials such as school principals.

But these villages have lost many of their youngest members to the allure of Bangkok—or overseas work assignments in the Middle East or Chinese countries. Does each household yearn for their own phone in order to keep in touch with their loved ones? No. The few cell phone owners are quite happy to rent access to their phones—often for a nominal rate of just US10c to US20c—to receive calls from further afield. Without this source of additional income, it is unlikely that many of them could afford the phone.

In Phimai's town center, the busiest place on a weekday afternoon is the wet market, followed by the local Internet café. With about ten terminals, it isn't cheap to use for a Thai teenager—at about US$1.50 an hour. But Thais are gregarious and typically each terminal will be shared by several friends who indulge in group messaging with their peers. The real economic activity remains down at the wet market—where the parents earn the real incomes that trickle down to their children's Internet educations.

Is Phimai a telecom-deprived place? Using a Western metrix, it would seem so. But the reality is that access is available for those who need it, at an economically sustainable pace.

The economic reality of shared—or public—access can be seen elsewhere in the developing world. Travel through some of the richer areas of Jakarta such as Menteng and Pondok Indah and you will see graceful Dutch homes with full access to broadband, cable television and basic telephony. Elsewhere in the country, the *wartel* rules. These are entrepeneurial small firms that offer the public a range of postal, computer and telecom services for cheap prices. Not many Indonesians have telephones in their homes—or in their *kampungs* for that matter—but few lack access to a neighborhood *wartel.*

Developed inhibitions

And so it is in the developed world. Like it or not, the most common form of private Internet access comes via a 56k modem connection that costs less than US$25 per month. DSL and cable modem services have been most successful when they have been priced—as a loss leader—to that level.

The multi-megabit connection is yet to become a standard business requirement. The few empirical surveys that exist of global telecom use show that the vast majority is still content with ISDN-speed connections.

Businesses don't necessarily want to furnish their employees with broadband access to the outside world.

Intranet access to the LAN server can usually be achieved via an Ethernet cable that runs inside the office. Many corporations and organizations seek to restrict Internet access and usage for productivity reasons. Moves to web-based email applications and other bandwidth-consuming activities often lead to reduced performance and productivity loss.

Cable modem providers inadvertently reinforce these attitudes with the restrictions they place on customer broadband use. The response of many university backbones to broadband-hungry applications such as Napster was to ban access to them.

What problem?

Too many in the industry see such phenomena as a mere blip—a short-term problem to solve and a long-term economic opportunity.

The reality is all together different. The telecom stock bubble occurred for the same reason that there was a railway stock bubble last century. Investors in both eras shared the same casino mentality. When disruptive industries come along that so obviously transform economic and social behavior—railways being one, the Internet another—many people assume that the normal rules of investment and economics are suspended.

But these rules tend to manifest themselves in annoyingly predictable ways. In the United States, more people own telecommunications stocks than subscribe to broadband services. Quite literally, the demand for the profits from telecom outweighed the demand for its few high-growth services. Telecom was hot with consumers because of competition, and the perception of falling prices and increased innovation. It was hotter with investors because of the historical reality of high margins and the hope that falling margins would be outweighed by the rise of volumes brought about by increased Internet and voice usage.

There were too many distortions in the market. Governments continued to make laws based not on equality of opportunity, but the pursuit of ideological or political ends. Regulatory dispensations to special interests distorted business cases and created illusory opportunities. In developing countries, investment opportunities were weighted in favor of local companies or those who benefited from political patronage. Even under regulatory regimes with good intentions, too many concessions and licenses were handed out with unreasonable costs or service requirements attached.

Technology disruption didn't help. In an industry with so many technology options, it was comparatively easy to make a case that a new or differentiated approach is worth a financial bet. LMDS looked sexier than DSL. IP telephony seemed more promising than callback. Freespace optics promised more than fiber. And with the rate of innovation increasing, so will tomorrow's promising technologies be undermined by even more exciting developments.

Telecom operators, with their insular mindsets and techno-centric views, often forget the little things. A buyer of telecommunications for an

Australian regional bank told me once of the reason he awarded a communications tender to a small start-up instead of its global behemoth rivals. The start-up had promised to page him personally if any of the data links to the bank's teller machines suffered an outage. The big boys thought this was an irrelevant detail in his Request for Proposal.

It's these types of details that will make a difference for the telecom industry. Some operators such as Equant and IXNet understand this and have reaped the benefits. But too many others are content to sit lazily at certain layers of the industry chain and divide their customers up by account size—rather than focusing on demonstrated behavior and tailoring their internal structures and external offers to fit.

Telecom operators—and their suppliers and financiers—have learned some billion dollar lessons over the past few years.

It's no good trying to push HFC-based pay sports broadcasts on households if the resulting content wars decimate the community's favorite football teams.

It's no good selling anyone a cable modem service if you then stop them doing things they want to do, like interactive gaming, web site hosting or video streaming.

It's no good putting in an eight-terabit cable to a country with a telephone penetration of less than three percent and negligible Internet penetration.

It's no good subsidizing the creation of digital slums—regions with barely-used broadband networks, but under-funded energy, transport and education systems and recessed economies.

Consumers want services that are cheap and which work. Too many telecom executives and investment dollars have chased an all-together different goal.

The big question. Have they learned from these lessons?

Epilogue

As this book was in the process of editing and publishing, the bad news continued to flow in. One of the few bright spots of a generally depressed telecom sector—DoCoMo's I-Mode service—was the latest to fall victim to irrational exuberance.

Keen to claim a lead as the first 3G operator in the world, DoCoMo had taken out millions of dollars worth of advertising in the world's business press touting its planned May 30 launch of high-speed W-CDMA services.

Then in late April, came the announcement that a commercial launch would be delayed until October—with only a small trial to take place in the interim. Investors wiped nearly $10 billion off DoCoMo's value on the day after the announcement.

Meanwhile, BT was hoping that it would be first to market with a 3G network, planned for deployment on the Isle of Man. One problem, though.

The network had just three handsets at its disposal and a shipment of several hundred more could be delayed for as long as a year.

The bubble bust was to continue on…

Glossary

3G: The abbreviated collective term for next-generation wireless standards that will support mobile voice, data and video.

911: The US emergency services number.

ASP (Application Services Provider): A company that provides value-added services over Internet connections

ATM (Asynchronous Transfer Mode): An advanced technology for transmitting voice, data and video popular in the nineties, now becoming superseded by pure IP and optical systems.

BOC (Bell Operating Companies): Collective term describing US' dominant four local operators—SBC, Verizon, Qwest and Bellsouth.

Broadband: A term describing a family of technologies that typically provide data rates of over 1.5 Mbps

Cable modems: Technology that enables multi-megabit data transmission over cable television networks.

CAN Customer Access Network

Carriers: Companies that literally carry telecommunications transmissions.

CDMA (Code Division Multiple Access): American-originated digital wireless air interface. Used in second generation cellular networks and likely to form basis for most third-generation networks. Patented by San Diego firm Qualcomm.

CDMA2000/1X: Collective terms used to describe Qualcomm-backed third generation wireless platforms. Most likely to be deployed by existing second generation CDMA operators.

CLEC (Competitive Local Exchange Carrier): A North American regulatory term used to describe companies that compete against Bell local service monopolies. There were about 400 CLECs in 2000 but their numbers have since dwindled.

DSL (Digital Subscriber Line): Technology that enables multi-megabit data transmission over copper telephone lines. Variants are called ADSL, HDSL and VDSL.

Exchange: Also called switch, central office. Equipment which routes calls and other telecom transmissions to their direct destinations.

Frame Relay: An established data technology which operates at between ISDN and DSL speeds.

Free-Space Optics: A fiber optic system minus the fiber, employing laser through the air to transmit data.

GPS (Global Positioning System): A US-developed satellite system which can pinpoint locations.

GSM (Global System for Mobile): European-specified digital cellular technology that is the dominant second-generation platform in the world.

HFC (Hybrid fiber-coax): A combination of cable technologies used to provide television, Internet and voice services.

IDD (International Direct Dial): Automated international telephone service.

IP (Internet Protocol): The system by which Internet data packets are addressed and routed.

ISDN (Integrated Services Digital Network): An historical forerunner of DSL services operating at speeds of 64-128 kbps.

ISP (Internet service provider): Companies which provide access to the Internet.

ITU (International Telecommunications Union); United Nations agency charged with setting global telecommunications standards and promoting development. Based in Geneva and funded by members according to economic size.

LAS-CDMA (Large Area Synchronous CDMA): A third-generation wireless standard backed by US-China venture LinkAir.

LMDS(Local Multipoint Distribution Service): A microwave-based technology that supports high-speed data transmissions.

MP3: An audio file format popular for peer-to-peer exchange of music and video.

Multiplexer: Equipment that combines multiple signals over one channel.

OFDM (Orthogonal Frequency Division Multiplexing): An emerging wireless platform some have tagged 4G as a result of its exponential improvement in data speeds.

PCS (Personal Communications Services): A North American term that describes cellular services operating in the 1800-1900 MHz band. Called DCS in Europe.

Pre-selection: The process that enables a customer to choose and then use a carrier without dialing a special code.

PSTN: Public telephone network

PTT (Post, Telegraphs & Telephone): A term that describes traditional government monopolies charged with providing communications services.

Resellers: Companies that market telecom services provided by other companies. Typically do not own infrastructure.

TDMA (Time Division Multiple Access): A second generation wireless interface, used by GSM, Japan's PDC and by operators across the Americas under its own name.

TD-SCDMA (Time Division Synchronous CDMA): A third-generation wireless standard backed by Siemens and China.

W-CDMA (Wideband CDMA): A European and Japanese-backed third generation wireless standard designed to provide upgrade path for GSM operators.

WTO (World Trade Organization): Global body of trading economies. Monitors adherence by members to free trade treaties, including telecommunications access commitments.

Primary sources for this book

This book is mainly based on interviews and press conferences with the following industry figures, compiled between 1993 and the present.

These include:

SingTel CEO Lee Hsien Yang

Former Hongkong Telecom CEO Linus Cheung

BT CEO Sir Peter Bonfield

Former Sprint PCS CEO Andrew Sukawaty

ITU Secretary-General Yoshio Utsumi

AT&T Asia Pacific VP Phil Overmyer

IBasis CEO Ofer Gneezy

Former ATUG MD Alan Horsley

Former Austel chairman Neil Tuckwell

Former TUANZ chairman Grant Forsyth

Former OFTA chairman Alex Arena

Former Pacific Internet CEO Nicholas Lee

Huawei vice-president Dr Zhijun Xu

BT vice-president Martin O'Connor

Former Worldpartners' president Simon Krieger

CDMA Development Group CEO Perry La Forge

Ericsson executive Larry Wood

Former AAP Telecommunications CEO Larry Williams

Former AAP Telecommunications regulatory affairs manager Brian Perkins

Former WorldXchange Australia CEO Richard Vincent

Former Australian Communications Minister Michael Lee

ATUG chairman George Maltby

Marc Faber

Former Worldcom director Colin Williams

Level 3 Asia CEO Steve Liddell

Other executives with Global One, Infonet, IXNet, Global Crossing, Worldcom, BT, MCI, Leap Wireless, Nokia, Lucent, Ericsson, PT Natrindo, AceS, Shinsat, IDA Singapore, Zhongxing Telecom, Marconi, Bosch, Swisscom, Telegroup, Vodafone, Universal Access, Bellsouth, Hutchison, Cignal, Extant, Arbinet, RateXChange, Asia Capacity Exchange, Indonesia Telkom, Alcatel, Mongolia Telecom, Pacific Internet, One.Tel, Concert, Globalstar, Orbcomm, Telephone Organisation of Thailand, Linkair, Qualcomm and Cisco.

Other sources

Among the editorial sources used for this book are:

—The authors' own articles in Communications Day, Telecom Asia, Wireless Asia and America's Network.

—Wall Street Journal, Financial Times, Red Herring, Upside, Industry Standard, South China Morning Post, Jakarta Post, Bangkok Post, Business Week, Asiaweek, Gilder Technology Report, Far Eastern Economic Review, The Economist, webb-site.com, TechBuddha, internet.com, Hong Kong I-Mail, Communications Daily and various analyst reports, SEC filings and message board posts.

—Conversations with contacts in operators, vendors and users groups in Hong Kong, Indonesia, Australia, the US, the UK, Germany, France, Sweden, South Africa, Vanuatu, Singapore, Malaysia, Thailand, Taiwan, Japan, Switzerland, China, the Netherlands, Kenya, the United Arab Emirates, Canada, India, Nepal and Finland.

Among the books consulted for this work are:
Michael Backman, Asian Eclipse, Wiley & Sons, 1999
George Gilder, Telecosm, Free Press, 2000
Reed Hundt, You Say You Want A Revolution, Yale, 2000
Howard Kurtz, The Fortune Tellers, Free Press, 2000
William Lee, Essentials of Wireless Communications, McGraw Hill, 2001
Michael Mandel, The Coming Internet Depression, Basic Books, 2000
Robert Schiller, Irrational Exuberance, Princeton, 2000
Nury Vittachi, Riding the Millenial Storm, Wiley & Sons, 1998
Michael Wolff, Burn Rate, Simon & Schuster, 1998

Where possible in the text, all sources for empirical data have been identified.

For more information and updates on the telecom industry

Grahame Lynch and his editorial team produce a number of daily, weekly and monthly newsletters that cover the telecommunications industry.

Based in Sydney, Australia and Bangkok, Thailand, with correspondents in India, Hong Kong, the United States and Britain, these newsletters are available for subscription at *www.decisive.com.au* and *www.commsday.com*

The newsletters include Communications Day, C&M and Tele 2.1

Archives of Grahame Lynch's articles are available at www.hive4telecom.com, *www.decisive.com.au* and www.telecomasia.net

Other good sources of news about telecommunications include:

The Wall Street Journal *www.wsj.com*

The Financial Times *www.ft.com*

The South China Morning Post *www.scmp.com*

Total Telecom *www.totaltele.com*

TelecomClick *www.telecomclick.com*

Teledotcom *www.teledotcom.com*

Teledotcom Asia *www.asiatele.com*

The Australian Financial Review *www.afr.com.au*

Internet.com *www.internet.com*

Index

CLECs, 5, 27-29, 31, 91, 96, 115
Clinton, Bill, 28
Coca Cola, 89
COLT, 3
Communications Day, 43, 121-122, 133
Concert, 87-88, 120
Covad, 28
Continental Cablevision, 18
Crowe, James, 1, 3, 51
DavNet, 21, 33
De Beers', 60
Deutsche Telekom, 87-88
DoCoMo, 31, 45, 106, 113
Doom, Gloom and Boom Report, 58
DSL, 4-5, 28, 30-31, 99, 106, 108, 110-111, 116
Engebretson, Joan, 27
Epoch Partners, 9
Equant, 89-90, 112
Ericsson, 43-45, 47, 49, 68, 72-73, 81, 103, 119-120
European Union, 68
Faber, Marc, 57-58, 119
Fanning, Shawn, 33
Far Eastern Economic Review, 10, 121
Fast Company, 33
Financial Times, 96, 121-122
Fischer-Madsen, Allan, 86
FLAG, 2, 8, 11-12, 94, 104
Fletcher, Daniel, 92
Forbes, 51
Fox, 18
Foxtel, 19, 21
France Telecom, 78, 87-88, 90

About the author

Grahame Lynch is a world leading telecommunications journalist and commentator.

He is the owner and director of Decisive Publishing, headquartered in Sydney, Australia, which publishes a range of regional telecom newsletters and web products including Communications Day. He is also the author of the Telecom Planet column, which appears in America's Network, Telecom Asia and Wireless Asia.

He was previously the group editorial director of Advanstar Telecom Group, with responsibility for the company's North American, Latin American and Asian magazines. In this position, he was based in Los Angeles, and prior to that, Hong Kong, China.

His writing has also appeared in The Euronet, TelePress Latino America, Telecom China, Australian Communications, Australian Personal Computer, The Bulletin, Business Review Weekly, Telenews Asia and The Australian newspaper.

Lynch has reported on telecommunications from 22 countries in Europe, North America, Asia and Africa and has interviewed hundreds of industry figures since 1993. His writing has appeared in three languages in over 100 countries and has been read by an estimated 400,000 telecom industry executives.

He is based in Bangkok, Thailand with his wife, Chongko.

www.ingramcontent.com/pod-product-compliance
Lightning Source LLC
Chambersburg PA
CBHW030755180526
45163CB00003B/1040